Marrow
Memory

Also by Margaret Nowaczyk

Chasing Zebras: A Memoir of Genetics, Mental Health and Writing
Polish(ed): Poland Rooted in Canadian Fiction (co-edited with Kasia
  Jaronczyk)
Searching for Ancestors (in Polish)
Your Family Tree of Health (in Polish)

Marrow Memory

ESSAYS OF DISCOVERY

Margaret Nowaczyk

Published by James Street North Books
an imprint of Wolsak and Wynn Publishers
280 James Street North
Hamilton, ON L8R2L3
www.wolsakandwynn.ca

Editor: Noelle Allen | Copy editor: Megan Beadle
Cover and interior design: Jennifer Rawlinson
Cover image: Benjamin Toth/iStockPhoto.com
Author photograph: Melanie Gordon Photography
Typeset in Adobe Caslon Pro, Veneer and Good Karma
Printed by Rapido Books, Montreal, Canada

10 9 8 7 6 5 4 3 2 1

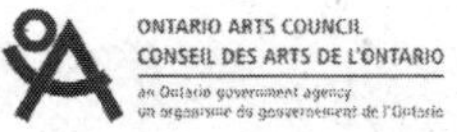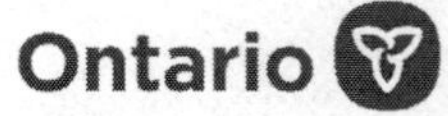

The publisher gratefully acknowledges the support of the Canada Council for the Arts and the Ontario Arts Council. We also acknowledge the financial support of the Government of Canada through the Canada Book Fund and the Government of Ontario through the Ontario Book Publishing Tax Credit and Ontario Creates.

Library and Archives Canada Cataloguing in Publication

Title: Marrow memory : essays of discovery / Margaret Nowaczyk.
Other titles: Marrow memory (Compilation)
Names: Nowaczyk, Małgorzata (Małgorzata J. M.), author.
Identifiers: Canadiana 20240361830 | ISBN 9781989496909 (softcover)
Subjects: LCSH: Nowaczyk, Małgorzata (Małgorzata J. M.) | LCGFT: Essays. |
    LCGFT: Autobiographies.
Classification: LCC PS8627.O9248 M37 2024 | DDC C814/.6—dc23

CONTENTS

PREFACE

In physics, there exists a concept of the resultant – a force that is equivalent to two or more forces acting on an object at the same time. As such, the resultant comprises all of its components in one single entity. This collection of essays is such a resultant: a composite of my preoccupations, obsessions, influences. My fascination with the mystery of memory, how we react to and interpret it, and its underlying molecular changes, and with the border between the biochemical and physiological workings of the body and its environment, the skin of existence. My preoccupation with the naming of natural phenomena in unrelated languages and with the feminist intersection of the personal and the political. My delight in being transported by a painting, how the mere action of jotting down a thought can free and heal, how sharing our writing empowers us.

Several of these pieces were written during my graduate studies as composition exercises and some afterwards, but all were penned in response to my inner inklings. They reflect the unadulterated workings of my mind, the marrow of my being. The bone marrow, this spongy, cellular matter that fills the tiniest recesses of our bones, is the innermost tissue in our bodies: it develops from the middle layer of the three-layered embryo. One cannot go any deeper than that. As such, it is an apt metaphor for our deepest self and it is from there that I wrote these essays. Several pieces in the collection deal specifically with my professional life as a physician and a scientist but never in isolation from who and what I am: a woman, a mother, a wife, an emigrant and a suffering mind. Tightly interwoven and interconnected like the cells of the bone marrow, these motifs blend into a cohesive, living whole. I could not hug my newborn sons without being astonished by the miracles of genetics and biology that brought them into my arms, or visit a remote hamlet in southeastern Poland where my mother's ancestors lived and not think about epigenetics and the inheritance of memories, any more than I could recollect a radio play I heard in childhood without marvelling at the miracle of hearing. Genetics and medicine are the underpinnings of my life, experiences and memories are what give it meaning and sense.

YOU, IN TRANSLATION

The poet moves from life to language, the translator moves from language to life; both, like the immigrant, try to identify the invisible, what's between the lines, the mysterious implications.
– Anne Michaels, *Fugitive Pieces*

Translate (physics) – to cause a body to move so that all its parts travel in the same direction, without rotation or change of shape.
– *Oxford English Dictionary*

You list your family's needs for the first week in Toronto: subway tickets, bed linens, child care for your six-year-old sister, certified translations of your father's university diploma and of your high school report cards. The Polish interpreter at the Ontario Welcome House utters four rapid words you don't catch to your caseworker. A crash course on the difference between an interpreter and a translator.

Your immigration caseworker snaps: "*We* say 'please' at the *end* of the sentence" just as you say, "Please, would you . . ." Her kilt swirls about her knees as she turns and sashays away and leaves you standing, stunned, in the middle of the waiting room at the Ontario Welcome House.

You instruct your new classmates and teachers how to pronounce your Polish name – maw-goh-JAH-tah. They mangle it and say: "It's beautiful." After a year of suffering their garbled utterances, you begin to introduce yourself as Margaret. No more exotica – you want to blend in, to belong, in name if not in anything else.

While discussing *Heart of Darkness*, where she disapproves of Conrad's usage of "moustaches," Miss Anglin, your grade twelve English teacher, announces that immigrants may become bilingual after many years but that their English will never be idiomatic. You're the only non-English speaker in the class – of course you take it personally.

You miss hearing people laugh at those one-line zingers you were so famous for.

After you get a "marginal pass" on the written English proficiency exam required for university admission, an alumnus of a Toronto private school that even you have heard of can't stop laughing, embarrassed – he got the same mark.

"Omit needless words," Strunk & White admonish. But you need to know these words first, you think dejectedly.

You stun your brilliant Canadian boyfriend by asking whether "lousy" comes from "louse" – it never occurred to him. You have to drag him to the library copy of the *Oxford English Dictionary* to prove to him that "a lot" is two words, not one: he wouldn't take your word for it.

When you read the journal you write in English, in your head it sounds as if filtered through a mouthful of pebbles.

A Canadian guy at the Harbour Castle bar tries to pick you up and tells you that you speak English really *good*. The irony is not lost on you, even if your accent gives you away, again.

You smirk after reading "The Owl and the Pussy-Cat"; it's cute, precious. You read the Polish translation and you burst out laughing – it's funny! A lightning punch of hilarity in your gut. In your native tongue, "a hog 'cross a road" evokes emotions and memories where "and there in a wood a Piggy-wig stood" fails.

When you meet the man you will marry, and whom you will bear two sons, he introduces himself as James. Like you, he has changed the original Greek name he was given at birth. You never call him Demetrios, and he never calls you Małgorzata.

For a long while now you have been unable to indulge in puns or neologisms – Canadians correct you because your accent contradicts your grasp of the language. When you go back to Poland after twenty-one years you aren't allowed to do it there, either – you are corrected because Poles think you've forgotten your native tongue.

You write and publish a book in Polish, a bestseller, but why does it feel like a step back?

When you travel to Poland you sometimes speak Polish to your sons and then insist that you've spoken English. They shake their heads, no, you haven't. They don't understand Polish, or Greek for that matter: your language of love and child-rearing has always been English.

You read a professional English translation of your book – you don't recognize the voice. But in a short story you write in English, your voice finally sings.

"Do you still speak Polish?" a venerable colleague asks in his strong German accent. "I'm bilingual," you answer after a moment's hesitation. After twenty-three years, finally, the correct word to describe yourself.

MARROW MEMORY

Jack and Luke wade through the tangled shrubs, careful not to tread on the sunken graves of the Old Brusno cemetery. Sandstone crosses appear within the underbrush, weather-beaten. Pale angels and ashen, sad-faced Madonnas peek between the tree trunks, haloed with blossoms of reddish-orange moss. Footed in the greenery of the underbrush, they seem to float. A world secreted in the deep shadows of an old-growth beech forest in the southeastern corner of Poland.

From the time our plane took off from Toronto Pearson International Airport five days earlier, my sons had complained incessantly: about the *bigos* they would not touch (they gag at the smell of sauerkraut, the fragrance of my childhood); about my speaking Polish to them ("Excuse me, but I know what language I'm speaking."; "Not here, you don't."); about the lousy cellphone coverage. Whenever I try to shoot a photo of either of them, they pull faces or swat me away. On the first leg of a month-long tour of my homeland, we are visiting Horyniec, a town on the Polish-Ukrainian border where my grandmother was born in 1917.

"Not another cemetery," ten-year-old Luke moaned when I told them where we were going this morning.

Jack scowled as only a teenager can: "You're crazy, Mama."

Perhaps I am. I was bitten by a genealogy bug when, on a microfilm at the Hamilton Family History Library, I located my grandfather's name in the 1908 baptism book from the now extinct Polish parish in Toporów. At that moment, I saw my grandfather – a tall, sinewy carpenter who had given me rides on his broad shoulders during Sunday walks – as a helpless, two-day-old newborn held over the baptismal font by his godparents. The distances and borders between me and that moment – the ninety-four years, the thousands of kilometres over two continents and one ocean, the three official languages: Polish, Latin and English – disappeared. An unbroken chain led from that newborn to the newborn Luke I had recently brought home. From then on, I dug deeper, further into the past. It was not only the thrill of adding new names to the family tree that drove the search. Something more powerful, something unnameable tugged at me across the chasms of time.

I had wanted to visit this cemetery from the second its photograph appeared on my computer screen nine years earlier, white crosses glowing among the misty green beeches. At the time, I had hit a brick wall in my genealogical research: my mother had no idea where her mother, my maternal grandmother, was born. My grandmother had died when my mother was eight and the fragile chain of memories had broken. As far as I knew, all my ancestors were peasants, and, unlike nobility, their family trees have very shallow roots: serfs were too busy struggling to survive to pay attention to their family history. My mother's ancestors had lived in the southeastern Polish borderlands for centuries. Indentured to the local nobles, the land serfs were passed like cattle from one landowner to another as the estates were sold. Until the early 1700s, they had no surnames. I had no idea how to look for them.

Then, on a dreary, rainy afternoon in March 2002, a flimsy airmail envelope delivered my grandmother's death certificate. On it, in bright blue ballpoint ink, "Horyniec" was scrawled as her birthplace. From that scribble by an unknown priest, I traced eleven generations of my mother's family back to 1775. For three years on Saturday mornings, I pored over grainy microfilms of baptism, marriage and death registers in Hamilton's Family History Library. I cheered out loud, fist pumping the air, whenever I bridged another generation. That drive did not disappear when I reached the end of the written records, in fact, it became stronger. I had to see the place my ancestors had called home. I had to walk this land. But something else lurked in my limbic system, in memories and in conjectures.

As a five-year-old, I had overheard a story about my great-aunt's father having been sawn in half by a band of marauding Ukrainians after the Second World War in a barn not far from Horyniec. In primary school, I had learned about the slaughter of Jews that had taken place among these trees – stripped naked, forced to dig their own graves, they were shot as they stood next to the holes in the ground – and about the cattle cars that had carried them to the Bełżec gas chambers, thirty kilometres north of this cemetery. For centuries before that, Polish manor lords mercilessly crushed the frequent Ukrainian peasant uprisings of these lands. Blood has bathed the roots of these trees; the ashes of the dead nourished their budding leaves. Is that why I had always been afraid of forests?

This morning we hike across the meadows, the thorny blackberry vines grabbing at my ankles. I wear sandals and soon my skin is shredded, bleeding, and I cannot but think how my blood might mingle with the blood spilled in these lands. But when I finally walk on the soft soil beneath the cemetery trees, I no longer feel fear. The smooth-barked beeches stand tall and straight like sentinels. Spirits gather around me. They are my blood and bone, too; they are my roots. I inhale the green scent of foliage and the musk of leaf litter under my feet. My pulse slows down, tension seeps from my body and drains into the soil.

A shadow of a stone foundation in the ground, overgrown with thick moss – remains of the little wooden Orthodox church that once stood here. I crouch down beside the nearest obelisk and crumble the black soil, the *czarnoziem*, so thick and rich it feels oily between my fingers. I bring it to my face – it smells of life, not death.

"Hey, Jack!" Luke calls out. "Look, a skull!"

Gods! Bones must have surfaced and of course those two imps would stumble onto them. I fear they will be traumatized for life. But Luke is pointing to a crudely chiselled skull grinning its teeth from the junction of a sandstone cross. My fingertips trace its Cyrillic letters: *Here lies the body that is the corpse of Ivan Pavlycho.* I chuckle at the morbid specificity of the inscription. Ivan died on January 12, 1898.

On the way back, we lose our path in the meadows: the spring has been unseasonably rainy and the grasses all but obliterate the narrow trail. Jack and Luke throw themselves into the cattails and fescue, shouting with glee when I cannot see them. Two puppies off the leash, high on fresh air and freedom. No trauma in sight.

In Polish, *mała ojczyzna* – "little homeland" – refers to where you are from, not necessarily in the physical sense, but where you feel most at home. I am shocked to realize that Horyniec is mine when two red-stockinged storks lope through the flooded meadows as if in a scene from my childhood fairy tales. I have coloured those long beaks and legs with crimson crayon, traced those enormous black-tipped wings spread in flight; only the squirming green frogs are missing from their beaks. The fragrance of the hay drying in heaps as tall as a house wafts by and a sensation arises somewhere between my gut and my heart – I *know* this meadow, this brook, these storks. But how can I? I never knew this village existed. I am a skeptic, a scientist; there has to be a logical explanation. Have I inherited awareness of this place from my ancestors?

Research has shown that DNA, the stuff genes are made of, can be modified as a result of lived experiences. Memories can be written not in the genes themselves, but *onto* them, as epigenetic changes. Tiny one-carbon methyl molecules attach to the DNA chain and influence the three-dimensional coiling of the double helix. Like barnacles that cover the rungs of a ship's ladder and prevent you from getting purchase – the ladder is still there, its basic structure unchanged, but you cannot use it without scraping the shells off first. In this way, acquired elements may affect inheritance and nurture may affect nature. Such epigenetic transmission across two generations has been documented in cases of obesity and heart disease. And trauma. Even responses to smell, at least in laboratory mice. The hormone cortisol released during periods of physiologic stress effects epigenetic modification: the methyl barnacles attach to the DNA and alter how genes can be accessed during a person's life. Is that why the whistle of a freight train filled me with dread as I lay in my safe bed late at night, hundreds of kilometres from where my ancestors' Jewish neighbours had been transferred to death camps? The English have a term for it: "what's bred in the bone." I call it marrow memory.

What about the memories of the sun slanting in mid-afternoon on a winter day in a Gliwice apartment or of ripe wheat fields around Horyniec, curling like waves in the sudden gust of an approaching storm? The mushroomy smell of the forest floor that I adore? Maybe the endorphins and oxytocin of happiness affect epigenetic processes the way cortisol of stress does. Bathed in happy hormones, my genes have been marked with the spirit of this place. The scientist in me likes that theory.

"It's kismet," the Prince says. He shakes his head slowly and wonders out loud why our paths haven't crossed before. "We are children of this land," he adds. But he was born in the same communal hospital in Gliwice as I was, four hundred kilometres west of here. He is a year my senior.

I'd met him earlier in the evening. I was excited to hear that the scion of the local aristocracy would be attending the monthly meeting of the local history club where I was giving a talk, and crushed when he was running late – I had never met a real noble. He lives in Warsaw, but visits Horyniec to attend to his family's affairs. When, after the meeting, he asked me to have a drink with him, I felt that I deserved it.

The Prince isn't really a prince – his grandmother married a lesser noble and as a result of this *mésalliance* his father did not inherit the title, but he is still a member of *szlachta*, the Polish landed gentry. In his mid-forties, over six feet tall, broad-shouldered, with a closely cropped fringe of black hair over his forehead, he bears a striking resemblance to King Jan III Sobieski, the Polish national hero who in 1683 defeated the Turkish army on the outskirts of Vienna and, as certain Poles like to remind the rest of Europe, saved Christendom. The only thing missing is the droopy moustache. He must realize how alike they look. I wonder if that is why he cuts his hair that way.

We sit at a table in the dining room of a small inn with hunting-lodge aspirations. The walls are festooned with heads of deer and boar, stuffed pheasants perch on the credenza, but it still smells of fresh plaster. It will take decades of guns hung on the walls and leather boots drying by the smouldering fireplace for the place to smell right.

The Prince has been regaling me with tales of his attempts to regain ownership of his family's properties confiscated after the war. "It's all a bit dodgy," he says, the last two syllables with an Oxford drawl. He huddles conspiratorially over the table. The Prince's grandparents, who owned the Horyniec estates, were forced by the Communists to relocate to a two-room apartment in Gliwice. Their palace was turned into a sanatorium for workers with arthritis, their arable lands and forests converted to a State Agricultural Farm, the watered-down Polish version of the Soviet *kolhoz*, imposed upon farmers after the end of the Second World War.

The innkeeper brings two tall, sweating glasses of *leżajskie*, a local beer. He bows as he sets them on the table and the Prince nods in silence. Manifestations of their marrow memories? Of the centuries-old feudal class system epigenetically encoded and internalized? Was that why I was excited when I learned that the Prince was coming to the meeting and why I had agreed to have a drink with him?

My childhood history lessons taught me that *szlachta* were bad. It was Communist propaganda, but as an adult I realized that a lot of it was true. Polish peasants were not emancipated until 1864 and when it finally happened it was to the great dismay of the Polish landowners. There were no schools for village children in nineteenth-century Poland. All my ancestors, up to my paternal great-grandfather, were illiterate. Had he been born two hundred years earlier, the Prince probably would have treated his peasants the way his ancestors did. He seems to have forgotten that the serfs who belonged to his ancestors were my ancestors. If it hadn't been for government scholarships my father – the youngest of six children of a minor farmer – would never have attended university in the 1960s, taken his family to Canada in the 1980s and made it possible for me to become a doctor and a professor. The Prince and his family contributed nothing to my education, yet now he wants to cajole free work from me by appealing to my local patriotism.

"It would be wonderful if you could write for the *Gazette*," he says. The *Horyniec Gazette*, his pet project, is a quarterly about local current affairs and local history, and one of his many attempts at modernizing the region. "We need the locals here – the ones who pooh-pooh what I'm trying to accomplish – to see that people like you have not forgotten whence they came. That you cherish your roots and Polish history no matter how far you have travelled, how accomplished and famous you are."

But during my many hours reading microfilms I realized that being an ethnic Pole or a Roman Catholic was not what mattered to me: I care about the chain of people who lived and loved and made children so that one day I could come along. Long dead and forgotten, now lifted from non-being by my dogged tenacity to find their names and the dates of their lives, they could've been Catholic Poles, Jews, Orthodox Ukrainians or Lutheran Germans: as long as there were blood ties that linked us, as long as I carried their genes, I didn't care.

I won't tell the Prince that, though. I want him to like me. Instead, I ask about his family archives.

"I can't count how many times I have seen volumes from Prince Poniński's library at international auctions," he says morosely. "Fifteenth-century folios, priceless. There is nothing left of the archives, either." Then he perks up, rubs his hands together. "You should be the director in charge of getting foreign subscriptions for the *Gazette*. I want to reach out to émigrés who left Horyniec. Just one subscription from the States would keep us afloat for months. Your vast genealogical experience would be invaluable here."

I suggest that he might be more successful selling subscriptions to the visitors of the local sanatoriums – Horyniec has been a spa town since the late 1900s. A local quarterly would be a nice way to keep in touch, I say.

"That won't work," he scoffs. "The head of the local government thinks I am trying to take over, usurp his authority. A typical example of post-Communist mentality. He's fomenting anti-elite sentiments in the village. You know how it is."

I don't, but as I sit here on his family's former lands where my ancestors were his family's serfs, I fantasize about being invited to his palace when he reclaims it. I imagine myself sitting by a ceiling-high stone fireplace sipping cognac from a snifter and discussing local affairs. Why not? The peasant molecules in my marrow enjoy his attention, even if the scientist ones in my brain know fully well that he only wants to use me. As his ancestors used mine.

But it's nice to have a beer with a prince, especially as I decline his offer of the *Gazette* directorship.

A few days later, I zigzagged the rental car between foot-deep potholes in a road that looked like it had not been repaired since the fall of Communism twenty years earlier. Luckily, there was no traffic, only an occasional horse-drawn wagon piled sky-high with hay.

"Mama, stop," Jack said suddenly.

I pulled over, almost into the ditch. "What's wrong?"

He wanted to go for a walk. The tall poplars that lined the road swayed in the breeze, their silver-bottomed leaves flittering like a school of fish. We ambled along the shoulder until we found a dirt path snaking into a field of wheat: it stretched to the green line of trees on the horizon. We skipped over the puddles in the ruts; it had rained heavily the night before and giant clouds were riding low, their tops brilliantly white in the blue sky that shimmered with heat, their underbellies leaden. The air smelled of wet hay and, faintly, of manure.

Halfway down the path Jack stopped. He inhaled deeply.

"Can you take a picture of me?" he asked.

What's that? He wanted a photo? I looked around: an unremarkable Polish countryside.

"Why here?" I asked.

He shrugged. "I like it here."

The sun at my back cast both our shadows on the green verge dotted with tall white flowers. His eyes scrunched up against the sun, his arms crossed over his chest in an awkward teenage stance, and yet my son looked like he belonged.

And, with a small tug in my chest, I realized that was exactly what I had been hoping for.

MATISSE, THE SEA AND ME

On an April afternoon fragrant with rain-damp chestnut leaves, I visited the Centre Pompidou in Paris. In the permanent exhibit on the fifth floor, I drifted from canvas to canvas exhilarated by the kaleidoscopic colours of Kandinsky's circles, Gauguin's bold palette, Cézanne's pure greens and oranges. But when I turned a corner between two rooms I froze as if splashed by a cresting wave. I had been plunged into an ocean.

White shapes, only vaguely reminiscent of live creatures but unquestionably marine, floated on an expanse of alternating azure and indigo rectangles. I recognized a shark and a squid's tentacles groping in the waves. A bird-shape – a frigate bird? an albatross? – soared in the middle of a kelp forest and yet – strangely – it appeared appropriately placed. A smoothly cut, continuous garland of floating seaweed fronds surrounded the edges of the panel. The larger shapes in the middle – the bird, the shark, the island of kelp – had cruder, more ragged edges that marked the continuous negative space of the checkered background. The two-metre-high, three-metre-long panel filled the pure white wall in front of me. I stared, motionless.

It was a place I had always wanted to visit – the South Seas. Suddenly, I was wading knee-deep in the warm pellucid waters, my toes squirming in the sugar-white sand, my calves tickled by schools of tiny fish, the sun beating on my back, coconut palms rustling over my head. Surrounded by a turquoise sea I floated and stared at the powder-blue sky as if peering into a shimmering bowl. My childhood dream come true.

In the right bottom corner of the panel, in flowing black letters: *Matisse 46*. Beside the blond wood frame, on a small white plaque: *Polynésie, la mer.*

I learned the word "lagoon" when the Wilson children's magic bed skidded on the sand of a South Seas island. Seven or eight years old, I was reading Mary Norton's *The Magic Bedknob*. The word "lagoon," *laguna* in Polish, enchanted me – it was foreign yet easy to spell and pronounce, but it was the descriptions of the landscape that hijacked my imagination and dreams for years to come: the coral forests in transparent waters, the anemones shivering their tentacles as fish darted in and out, the same yet different blues of the sky and the water, the whitest of sands.

Before I learned to read, I had entertained myself by poring over the thirteen Polish tomes of my parents' *Great Encyclopedia PWN*. My favourite images were the colour plates of coral reefs and deep-sea fauna, and of aquarium fishes. There was even a plate devoted to kelp and seaweed – green, brown and shades of yellow. I had traced my finger along diagrams of anemones, sea cucumbers, starfish, fishes of all shapes and sizes. There were colours unseen in my grey world: turquoise, celadon, cerulean, oranges and pinks of all hues. The waters of faraway lagoons lapped gently in my imagination. One couldn't travel any farther from Communist Poland than the South Pacific, both figuratively and literally: nothing was more devoid of colour than a Communist apartment building complex in the '70s. But on the other side of the globe iridescent violet fish darted between the fluttering tentacles of pink-orange anemones, while dainty crimson seahorses and yellow sea dragons glimmered among fingers of coral. A rainbow paradise beneath the impassive surface of the ocean. "Ocean." O-ce-an – the susurration of the waves captured in the word itself.

Other words also kept me in thrall: "Atoll." "Coral reef." "Lunar tides." "The Gulf Stream." "Starfish." "Mother-of-pearl." "Mollusks." "Parrotfish." And the names of the archipelagos and islands: Tahiti and Mooréa. Bora-Bora. Palau. Polynesia.

One early December dusk when I was five or six years old, I stood in front of a bookstore window lit from within like a huge aquarium. Right beside the glass pane was *A Stroll in the Depths*, a marine biology book with a cover photo of a pink-orange sea anemone and a striped hermit crab under the dome of blue waters – just like I had imagined an atoll would look. I scrunched my nose to the glass, my breath fogging it. The colours glowed magically amongst the grey propaganda pamphlets and school textbooks bound in dull pastels. I demanded my parents buy the book for me. I cajoled and wheedled, possibly threw a tantrum, and wouldn't stop begging until my father finally brought it home a few days later. I flipped through the pages of the slim volume, mesmerized by the pictures of jellyfish with see-through blood, microscopic diatoms and pink octopuses. Oh, to dive into waters that teemed with such creatures, to have the fish swim between my fingers. To see a real orange hermit crab living inside a spiral whelk shell topped with a violet-tinged anemone.

I wanted to go to Tahiti.

*

Matisse visited Tahiti in 1930 at the age of sixty. For years before the trip, he had felt creatively blocked. He believed that his brain and eyes had fallen into a creative rut, and only an infusion of new stimuli – light, colour and shapes – would shake him out.

After visiting New York City and Chicago he set off from San Francisco on March 20. The ship was RMS *Tahiti* – a battered old English mail boat with a surly captain, dismal food and the dull company of Australian businessmen and sheep traders. He revived only as the ship crossed the equator and the hue of the sea lightened from dark blue to the blue of the morpho butterfly – a hue that Matisse considered his talisman.

At first, he seemed irritated by the unchanging beauty of his new surroundings: "It was both superb and boring . . . the weather is beautiful at sunrise and it does not change until night," he wrote to his wife after disembarking in Papeete. "Such immutable happiness is tiring."[1] But he was also dazzled, especially by the light, which he compared to gazing into a golden goblet.

He travelled extensively around the island by car and was enchanted by the colours and the bounty of the local markets, the shades and hues and textures of the Tahitian flora, and the never-ending spectacle of the tropical sunsets. Two months into his stay, he sailed to the Fakarava Atoll – a strip of coral reef, about 270 metres wide, enclosing an inland sea 130 kilometres in diameter. Here, Matisse spent whole days swimming and diving in the lagoon, gazing into the translucent blue waters that shimmered over the banks of coral. He described the waters as grey-green jade shading to absinthe, peppermint green and blue.

He glided among fish "glinting like enamel or Chinese porcelain, minute, jewelled specks and streaks of colour swarming round

---

1   Hilary Spurling, *Matisse the Master: A Life of Henri Matisse: The Conquest of Colour, 1909–1954* (New York: Knopf, 2005).

his legs." Diving mask on, he dunked his head in and out of the waters, training his eyes on the different luminosities and saturations of water and sky. Years later, during the dark nights of World War II, when he reminisced about the atoll, he wrote in his diary of its waters, the soughing of the trade winds, the silk-like coconut palms and of "the overall impression it gave of power, youth, fullness and completion." He revered the austere simplicity of the outports: "the sky and sea, coconut palms and fish – that's all there is to see – with a pure radiance that makes them incomparably precious."[2]

But it would be almost a decade and a half before Polynesian fauna and flora and the sea itself flooded onto his canvases.

The first sea I experienced was the Baltic. I have a memory of holding onto somebody's sinewy back – my grandfather's? – as he swam, or maybe waded, in salty water. Pale beige dunes traced the horizon behind a wooden pier. Gulls cawed over waves topped with silvery foam. Far from tropical, the cold Baltic was more grey than blue – gunmetal blue. Dark gold sand was layered upon a bed of tiny pebbles about six inches below the surface. Those pebbles provided a perfect substrate for paving the courtyards of the sandcastles and the streets in the mud cities the kids erected, every single one surrounded by a moat. Brown, reddish, maroon, pearly white and yellow; rough and gritty between the child's fingertips digging wells and against the tiny palms smoothing out the stones. The beach cut a wide swath between the sea line and out-of-bounds dunes. Signs forbade entrance every fifty metres; the only access to the beach was by means of trails that led perpendicularly across the sandy ridge from the tall pine forests.

Few shells dotted the Baltic beach, all of them bivalves: bulbous mussels and thin tellins. Ivory-coloured, fluted ovals the size

---

2   Spurling, *Matisse the Master*.

of my thumbnail, with a pinkish hue on their inner surface. The prize find were two tellin shells still hinged together, like butterfly wings; the larger the better. But I wanted whelk shells with inner coils enamelled a vivid pink-orange. I wanted spiky sea urchins and crabs and sea anemones. Starfish and coral. Conch.

The Baltic offered jellyfish and amber. The Baltic moon jellies – translucent domes of quivering goo – swarmed the waves by the thousand every August; their four horseshoe-shaped gonads turning more and more brilliant as they reached maturity. Neon blue, vivid violet, hot pink. We caught them in our cupped hands and our mothers would yank our elbows, shouting, "You'll get stung!" as they beat them out of our grip, the poor creatures pulverized.

"They grow their bodies back," my father said when, in tears, I brought a tattered fragment to him. "You get more that way." I think he confused them with earthworms, but at the time I accepted this as an example of eternal life.

And amber. The gold of the sea. The petrified resin of the evergreen trees that blanketed the Baltic region forty-five million years ago. Children and adults alike scoured the pebbles along the beach in search of the buttery yellow and light orange nuggets: semi-transparent, they glowed with an inner fire when lifted to the sun. Lighter than stones, they had a smooth, almost yielding, surface that fingertips slid off. In the seaside museum I admired honey-hued pieces the size of my child's fist, a perfect beetle or fly preserved in their centres, but I never found any myself. I was either unlucky or not patient enough.

Pictorially speaking, Matisse returned empty-handed from his two-month sojourn in Tahiti. His entire output while there comprised one oil sketch and about twenty pen-and-ink drawings of views from his hotel room: coconut palms, lagoons and a few people. For almost fifteen years, before he transformed the walls of his Paris apartment with the designs for *Océanie, la mer* and *Océanie, le ciel*, the two designs that preceded *Polynésie, la mer*, Matisse's creative use of Tahitian memories was sporadic. The pen-and-ink drawings inspired illustrations for *Pasiphaë* and two Tahiti-themed sculptures. It was as if the radiance, the hues and the textures had to be digested by his subconscious, and only when metabolized into their primary elements and colours – what Matisse called "essences" – could they find their way into his art.

In 1945, Matisse was commissioned to design two tapestries for the famed Gobelins Manufactory. By this time, he had written to his daughter that he had gone as far as he could with oil painting. Visitors to his apartment found him seated in his wheelchair – after two surgeries during the Second World War, he was left with a massive abdominal hernia that made walking difficult – a huge pair of scissors in his hand, carving boldly into sheets of paper painted in bright primary colours. The switch to cutouts might have been dictated partly by his deteriorating health: standing by an easel was out of the question.

For *Océanie, la mer*, he cut white shapes from a block of writing paper, the only material available in Paris in 1946. Beginning with a paper swallow, he asked his night nurse to pin it over a stain on the wall. The apartment had not been redecorated for twenty years and was covered by a pale lining fabric that had darkened over time to a warm sandy brown. A fish followed. Whenever he suffered from insomnia, he would cut out a new shape and direct the nurse to pin it to the wall. The designs spread and spilled over the doorway, rounded the corner to flood the adjacent wall.

Marine life teemed in the room as if in an aquarium: loops and trailing fronds of seaweed surrounded swooping birds, dolphins, sharks, jellyfish, starfish. Elements were moved around, sometimes removed altogether, to achieve a balance, a final aesthetic that satisfied Matisse's sense of composition.

The cutouts, while appearing fragile and ethereal, were at the same time tangible, palpable. Once pinned onto the wall or the canvas they trembled with the breeze, shook with footsteps or the rolling of his wheelchair, as if buoyed by waves, stoked, fanned by a gentle breeze across an atoll.

Was Matisse remembering the Tahitian twilight during those sleepless nights in Paris? Maybe the colours of the tropics saturated his charcoal nights the way they filled my grey childhood days. Did he hear the breeze tousling the palm fronds and the waves lapping the coral reefs under the starry equatorial sky? Maybe the stiff paper reminded him of the rough edges of coral he fingered in Tahiti, the crunch of scissors like the crunch of shells under his feet on the beach. Memories rising up from the deep crevices of the subconscious like strings of air bubbles racing to the surface of the sea.

I can sit for hours watching waves pulsate on a beach. Rocky, sandy, pebbly, white, coral-pink or lava-black – it doesn't matter as long as the water is salty and the air tangy. And even though I do enjoy luxuriating on Caribbean sands, I prefer the colder beaches of the temperate zones – Tofino on Vancouver Island or Cape Ann in Massachusetts – where the sea is greyer, the waves higher, the breeze colder and saltier, the beige sand more earthy, more primal. The whisper of the waves induces a calm that eludes me in the city, that I search for in vain on a mountain hike or wading through waist-high meadow grasses. It is the combination of the sound and the scent of briny seawater – the iodine, other microelements – and the breeze that brings those scents into my olfactory bulbs to resonate in my limbic system, in the deep emotional pathways of the brain.

Why do I love the sea so much?

Does the lull of the waves calm just me or does it affect all human beings this way? Does it remind me of the sounds I heard while in my mother's womb? Swooshes of borborygmi in my mother's gut? Not rhythmic enough. Her heartbeat? Too fast. It must be her breathing I heard from inside her body. During prenatal life is the only time that we hear somebody else's breath from within. We can hear others' breaths transmitted through air that separates us, but it is only in the womb that our developing ears – the snail-like cochlea of the inner ear – are exposed to the incessant murmuring of our mothers' lungs. Solid tissues like the liver and the muscles of the diaphragm and the uterus transmit sound beautifully, faithfully, much better than air in the negative spaces between us and other humans after we are born. Air dulls and dissipates sounds. Sound transmitted through solid tissue has a lower timbre, lower frequencies; it is deeper, softer and yet more powerful, resonating in the fetus's bones and heart. Maybe our mothers' breaths resonate within our "gut brain" – the two layers of neurons filigreed throughout the wall of the digestive tract from the mouth to the anus. Does this continuous plexus of nervous tissue remember these primordial sounds, encoding our mothers' breathing into our beings? Do we find that rhythm again in the sounds of the waves?

And that wondrous internal sea of amniotic fluid, in which we float for forty weeks. The fluid that dulls the sounds of our mothers' bodies and the hubbub of life beyond their boundaries, that protects us from bumps and bounces of external life. Warm, buoyant, it lulls us; its smell fecund, luxuriant, earthy when it splashes onto the hospital room floor during labour. Seawater smells differently – of tangy, biting iodine salts and saltwater creatures' last days alive – but maybe in both we recognize the similar primal forces of nature and life. When I get off the plane in Hispaniola or Guadeloupe, when I walk the beach in Pacifica at sunset or on Naxos at sunrise, I fill my lungs with this smell, suck it into all the crevices of my body.

I wonder if Matisse felt that, sitting on the sandy slip of Fakarava Atoll.

The designs for *Océanie, la mer* and *Océanie, le ciel* were rejected by the Gobelins as being too difficult for their weavers. They claimed the sandy-brown background was impossible to match and Matisse began work on a pair of alternate designs. The new order requested a blue colour scheme. The turquoise and verdigris sheets his assistant found after scouring stationery stores in Paris became the checkerboard backgrounds for *Polynésie, la mer* and *Polynésie, le ciel*.

Was it the colour scheme of *Polynésie, la mer* that affected me that day in Paris? The calm background in alternating shades of blue carved into a continuous negative space by the white shapes? The white was not uniform; not all the life forms were cut from single sheets of paper – I noticed blotches of pale celadon, varied pinks and a coffee-stain beige in the larger shapes. For those, Matisse glued paper sheets together, and as they aged they acquired different hues, their whiteness becoming less stark, more natural.

In Matisse's later artwork the organic and aquatic images from his trip appeared in a series of "signs" – increasingly summary outlines of the objects, a distillation down to an "essence" of the object. Whatever he was working on – a leaf, a snail – he drew numerous times until the "essence" became apparent, until he was satisfied with its meaning. Only then did he pick up the scissors and free the image from the paper – the scissors held wide-open, carving, never clipping – through the sheet of pure colour. "Cutting straight into colour reminds me of the direct carving of the sculptor," he wrote.[3] The medium was more physical than painting – the resistance of the painted paper pushed against the shears.

On the *Polynésie, la mer* panels, to mark the ocean surface, he arranged the fronds of seaweed beneath long, thin rectangles of white paper. There were four such figures in the panel: three close to the top and the fourth – illogically – almost in the centre of the composition. Equally illogically, a fish hovered above a diving bird. The sea was vast, had many surfaces; space was warped in the ocean; life there existed on many levels. But the seaweed hung down vertically, not pulled by currents or battered by waves. The waters were calm and, in turn, I was soothed.

While my retinas feasted on the white shapes hovering amid turquoise and indigo waters my visual cortex processed neural impulses. Deep down inside the sulci of my limbic system they met my memories and, in my gut, they sparked a remembrance of smells and sounds – a near-physical experience.

Three years after I saw *Polynésie, la mer,* I snorkelled in a reef in the Caribbean archipelago of Turks and Caicos. After a lifetime of waiting, a quarter of a century after the word "lagoon" floated into my consciousness, I snapped on a diving mask and waded toward the coral reef that abutted the beach. Thirty metres from the shore, snorkel between my teeth, I dove in.

The silence surprised me: had I thought about it I would have known that there would be no sounds underwater. The rumble of the waves, the calls of the pelicans patrolling the beach disappeared the moment the shells of my ears submerged. Inside, the sea was serious, quiet, thoughtful. Sunlight, scattered by the surface waves, dappled the sandy bottom as if with a shimmering of coins.

---

3   Spurling, *Matisse the Master.*

Madrepores and brain coral, staghorns and tubes of pillar coral shadowed the sand in irregular splotches about three metres below me. As I floated, clouds of silvery fish the size of my little finger darted around me; larger solitary fish flew in and out of crevices: bright yellow and striped beige and tan, some with a brilliant double violet stripe, some pitch-black with bright green eyes. I spotted a green sea turtle. As it slowly paddled away, its underbelly a dull yellow, it resembled a graceful, if somewhat stout, angel.

But where were the hues of those old encyclopedia plates? The pinks, oranges and purples? The brilliant rubies and turquoises and emeralds? In real life, everything looked as if wrapped in a bluish-green mist. The coral clusters were various shades of grey, as if washed too many times, no red or pink in sight; if it weren't for the sun, they would have been the colours of the housing projects of my childhood. Were those old illustrations only an artist's vision? Did the brilliant colours of the underwater live only in my and Matisse's imaginations?

When I surfaced, the waters *were* jade and turquoise under the brilliant blue sky; the sand of the beach glared in the sun. Floating on my back with the gentle wind brushing my face I finally discovered the atolls and lagoons I had dreamed of in my childhood. Two days later, when I dove in the cerulean blue of the deeper waters on the other side of the island, I found an empty conch shell the size of a football. Its thick, wide, salmon-tinted lip, rough on the outside and smooth and glassy where it coiled into an internal staircase, invited my imagination to roam again.

AD INFINITUM

Once, as an intern, I pronounced four people dead in twenty-four hours; by four o'clock in the morning, I was ready to go home to my husband of six months, but I had to stay until the morning report. The sparrows were rioting outside my call room window that June morning.

Once, as a senior pediatric resident, I had to finger extract stone-hard feces from the rectum of a screaming five-year-old boy. The reeking stool clanged as it hit the metal bedpan.

Once, as a hospital volunteer, I vomited my lunch while watching a percutaneous needle biopsy of a coin lesion of the lung through a window in the fluoroscopy suite. I did not know what a biopsy was. Even if I had known about the lethal prognosis of a coin lesion, I wouldn't have cared: I only worried that throwing up in the hallway proved I wasn't meant to be a doctor.

Once, on a windy autumn day, I stood on the curb of a busy downtown street outside a teaching hospital and watched as doctors – I was sure it was doctors, it must have been doctors, who else could it be – talked a white-haired woman in a hospital gown down from the ledge of an eighth-floor window. Later, I would learn that that window belonged to the hospital library where I would spend nights and days studying.

Once, finally a medical student, I stood in awe in the operating room while pink-petalled flowers of human tissue blossomed in the mass of a collapsed grey-brown lung as it was reinflated with a hiss during a thoracotomy. It was magical. Later, at home, I wanted to write a poem about it, but I did not know how.

Once, while on duty in the emergency department, I watched as a seventeen-year-old girl's left pupil dilated and her left eye deviated down and outward, and I knew then and there that an artery inside her brain had just burst and was killing her. I wanted to crack open her skull to let the blood out, to allow the brainstem to spring back to its normal shape and consistency, but I called the cardiac arrest code instead and the cardiac-arrest team pushed me aside as the girl died in a cacophony of alarm bells.

Once, as a young attending, I was too exhausted to go to the hospital in the middle of the night because my colicky firstborn son's cries had kept me awake for ten months straight, and the patient, an eleven-year-old girl with advanced liver disease, suffered brain death before I saw her. At her funeral, the parents thanked me for the wonderful care that she had received at the hospital.

Once . . .

Once . . .

Once . . .

THIS IS RADIO WARSAW . . .

"This is Polish Radio One. It's twenty-two-hundred hours. In the news tonight . . ." The announcer's clear voice fades into static as I twist the dial. The display of the Grundig radio cassette player glows green in the dark and faintly illuminates the names of the European cities assigned to different wavelengths. I don't want to listen to the news, I'm searching for Radio Luxembourg, which is around here somewhere.

Around here is the 95.5 MHz range just below the Polish Radio One frequency. I need to twist the dial ever so slowly or I'll miss the faint, remote-sounding music. Radio Luxembourg plays the most recent hits from the West, its transmitters pointed to all points of the compass, including Eastern Europe. More often than not, I can't find it – the frequency is screechy, staticky or simply silent. Sometimes, wiggling the telescopic antenna topped with a bulbous knob helps. "Sunspots" is one explanation for the failed frequencies; "Communist radio jamming" the other, the latter more often than not directed at Radio Free Europe. But I am not looking for banned news about failing Communist regimes or the most recent human rights violations behind the Iron Curtain. I'm eleven years old, all I want is some rock and roll.

Here! A faint bass rhythm. I turn up the volume to catch the higher notes, a rapid dance beat with vibratos in the singers' voices. My feet slide across the floor of my room, which is not much larger than a closet (not that I know what a walk-in closet is), my shoulders twitch in tune with the rhythm of a faraway disco.

When the announcer's voice comes on, I turn the dial again. The red-striped frequency indicator passes the Polish Radio One bandwidth again and a chatter of distant voices and a siren Dopplering away pour out of the speaker. I stop mid-twirl.

"What's happening?" A lovely alto floats from the speaker as the chatter fades. The sound is clear and crisp, the words enunciated in the manner of a theatre actor.

"Somebody jumped," a man's voice replies. "From the building across the street." The woman gasps.

Must be radio theatre. I turn up the volume ever so slightly – I'm supposed to be asleep, at 22:41 it's well past my bedtime – and I listen.

```
Woman: How horrible. I can't see anything.
    The edge of my building blocks the view.
    I only see people's heads.
Man: I can see a white sheet on the ground
    from here. I think they left it to cover
    the blood.
W: (gasps) You never know what pain people
    are hiding. You could meet them in the
    hallway and never know.
M: Have we met?
W: I don't think so. My entrance is on
```

> Brzozowa, this is the back room.
> M: I'm on Zwycięstwa. A whole block away.
> W: So how can our windows be neighbours?
> M: It's these pre-war buildings, all corners
>   and dead ends.
> W: We could've passed each other down on the
>   street and never known we shared a wall.
> M: A wall, but not the view. I'm Piotr, by
>   the way.
> W: Anna. Nice to "meet" you.

The woman's lovely laugh fills my room. In the dark, I see their buildings clearly – tall, red brick sooted by car and factory emissions, with nooks and crannies, alcoves and balconies, pitched roofs punctuated by tiny attic windows. Few of those remained after Warsaw was demolished during the Second World War – of course the play took place in Warsaw, that's where all the interesting things happened. The apartments were centred around a well of a courtyard with hardly any sun reaching the cobblestones at the bottom. People's voices carried up, amplified, and everybody in the building knew everybody else's business.

In my room, the radio broadcast was my constant companion, turned on the moment I walked into the house from school, switched off only late at night, after I listened to it for hours already in bed, volume turned down low, low, hiding from my parents. I did my homework with its accompaniment, I ate my dinner at my desk when it delivered the news, I read while it sang and chattered.

I travelled the world when I turned the ribbed plastic knob – I hovered over the Danube bridges in Budapest, the columns of the Acropolis, the craggy peak of Mont Blanc. Depending on the frequency range I chose on the main selector, the static might yield only silences and one or two crystal-clear local stations (ultrashort waves), or a cacophony of languages and musical styles flowing seamlessly from one to the other (long waves). I learned to distinguish languages – Bulgarian, Portuguese, Swedish – as I passed the frequencies for Sofia, Lisbon or Stockholm. Czech sounded the funniest – many of its words sounded like childish diminutives of Polish ones; Russian was partly understandable, English less so, in spite of my private conversation classes. Sometimes, I managed to catch a Polish-language program from the West: BBC and Voice of America each broadcast hour-long newscasts in Polish and English language lessons, and, of course, Radio Free Europe, the latter almost never audible because of jamming. The ridiculous Radio Tirana broadcast inane anti-everybody – Soviets and Western powers equally – propaganda in every European language. This one I avoided, having once or twice fallen for its lure but quickly learning what rubbish it was.

Floating through the radio stations, the voices, languages and accents mingling freely bestowed a sense of freedom. The images of faraway and exotic lands mentioned on the news shimmered in my mind: the pyramids, New York's skyscrapers, Big Ben. As I listened, the world filled with the promise of travel and success – the night flower of my imagination blossoming in the darkness of my bedroom under the receiver's little green eye.

On the broadcast, the window squeaks open to usher street sounds; a car honks in the distance.

```
Anna: You there, Piotr? I can never tell.
Piotr: These buildings – keeping people apart
     as they crowd them together.
(birds chirping)
P: Hey, can you reach around the corner? Yes,
     I can see your arm!
A: (giggles) Now your turn!
(long silence)
```

```
P: I can't reciprocate, my arms aren't long
   enough. (silence) You see, I'm humpbacked;
   doctors call it kyphosis, but it's a hump
   alright. And I'm a dwarf, all short arms
   and legs.
A: It doesn't matter.
P: Oh, it does, believe me. You have no idea
   what it means to be different.
A: Don't I? I'm (pause) ugly. Really ugly. Not
   the jolie laide that is so desired for
   high fashion models, just plain ugly. At
   work, I just want to disappear under the
   floor when people talk to me. I can feel
   their stares, even on my back.
P: I'm sure it's not that bad.
(sounds of traffic for long time)
P: Are you there?
(sounds of nose being blown, sniffles)
P: I'm sorry, I didn't mean . . .
A: Good night.
(window shutting, glass panes rattling)
```

I see her as plainly as if I stood inside her room, her head veiled by the crocheted lace curtain, her silhouette black against the darkening sky. She rests her forehead against the windowpane for a moment or two – so lonely. She twists the handle of the window, shakes off the curtains and pulls the drapes across the window in one long swish, closing off the outside world.

At night, the radio signal was much crisper, cleaner. Bouncing toward the earth from the ionosphere and free of the sun's interference, radio waves travel farther at night. Nighttime was the time to listen to not only the Eastern European broadcasts, but to the stations from the West, the Middle East and northern Africa. Then I had the radio to myself, uninterrupted by my parents who might comment derisively about the news or my choice of music. It was only the broadcaster's voice and my eardrums resonating with it – nothing stood between us. As intimate as whispers in my ear.

During the day, I could only get the local, very-short bandwidth radio and the long-wave programming from Warsaw transmitted from the mast antenna in Konstantynów. Guyed by fifteen steel cables and 646.3 metres tall, it was the tallest structure in the world at the time. It stood southwest of Warsaw, its position an approximation of Poland's geographic centre, and it broadcast at AM-longwave 227 MHz frequency, which could be heard as far as Canada's North without satellite relays.

Radio waves are electromagnetic waves produced by the oscillation of electrons in the transmitting antenna. The longer the radio wave frequency, the farther it travels and more easily it penetrates obstacles in its way. Longwave radio can pass through buildings and foliage. As their wavelength increases, into very long and extralong frequencies, radio waves cover the entire globe and reach deep into the abysses of the ocean. As they travel through the ground and air, the electromagnetic waves can bend over hills and mountains to reach beyond the horizon: those surface waves wrap the entire globe in a vibrating cocoon, shot through with radio's invisible gossamer threads radiating from the broadcasting antennas.

The radio broadcast continues and I listen. I want to know what happens to these two.

> (street traffic sounds)
> P: What a beautiful sunset tonight.
> A: I can only see the sky overhead, mauve and pink. The corner blocks the west.
> P: The sun has just passed the rooftops, it's still a while before real dusk. But the streets are already shadowed. The day ends much earlier down below.

A: Can you see stars at night?
P: There's a soffit above my window and those
   tall building across the street. But I
   can sometimes catch the moon.
(distant sounds of traffic)
P: I've been thinking. Why don't we meet? You
   know, in person?
(long silence)
P: Anna?
A: I've got to go. I have an early day to-
   morrow.

I would love for them to meet. They seem so comfortable together. Maybe – my eleven-year-old self thinks – it wouldn't be so
bad for an ugly woman to marry a hunchback? These two might
never find anybody else.

As I listened, the electromagnetic waves received by the radio's
antenna vibrated the speaker membrane – with each arriving
electrical pulse, the membrane wrinkled and smoothened, like
the surface of the lake after a stone is dropped into it. The membrane's every move, every twitch produces pressure waves in the
air – sound waves – which fill the room. My ears, pink shells of
skin-covered cartilage, convey the pressure differences of the sound
waves into the ear canal, which ends about an inch into my skull at
the eardrum, which acts exactly as it name implies – as a drum. The
vibrations of this translucent, pearly-white membrane activate the
ossicles, the three tiniest bones in the body with the fanciful names
of "malleus," "incus" and "stapes": Latin for "hammer," "anvil" and
"stirrup." Together, the ossicles transmit the vibration of the eardrum to another drum-like membrane that covers the entrance to
the snail shell–shaped cochlea, a sensory structure seated deep in
the base of the skull – the inner ear. As this second membrane – the
oval window – vibrates, it produces waves in the fluid that fills the

cochlea. Tiny specialized neurons called hair cells, their sensors floating in the liquid, are triggered by the fluid waves and instantly fire impulses that travel up the auditory nerve to the brainstem and then to the brain's cortex just above the ears. There the real work of hearing ensues – meaning is attached to sounds and noises. This progression of information from sound waves to fluid waves translated through electrical impulses into conscious thought falls just this side of magical.

But this intertwined dance of physics and neural impulses does not explain the emotional responses to music or sound, what we experience when we hear a tune from our childhood. I can still hear Donna Summer's "Love to Love You Baby" as I heard it one sultry night of the summer of 1975, fragrant with nightstock that blossomed outside my window. That night, I could just see the New York City disco, the bright-coloured clothes, the sequined dresses shimmering with light. And it does not explain the feeling of be-longing I experienced while listening to the radio at night. Thou-sands – no, millions of people, ears turned to their little transistors or wood-cased table radios, all listening to the same words, music or drama. A community of souls, experiences and dreams, united by invisible waves, listening intently, hearing the same sounds, the same words, the same melodies. Connected. All so much more poignant for their lack of the visual, so dependent on imagination to supply the details. Alone-not-alone in my room, I drank in the voices, the sounds, my mind supplying the scenery, the costumes, the expressions on the faces of the actors.

In the broadcast, I hear the clinking bottles of the morning milk delivery and a honk of a bus: the sounds of a city waking up, birds chirping in the background.

        (window squeaks open)
        P: Anna? Are you there?

```
A: Good morning.
P: Have you thought about it?
A: About what?
P: Meeting in person. Let's do it! We talk
   so freely here, we'll have no problems in
   person.
   (silence, car horn in the distance)
A: (haltingly) Oh, okay.
P: Really? Great! Where should we meet? How
   about the old city clock? You know it?
A: In the park on Królewska?
P: That's the one. Saturday, seven p.m.?
A: How are we going to recognize each other?
P: (laughs) Let's do that old cliché - I'll
   carry a red rose. You do the same. I don't
   want to be accosting strangers.
A: (hesitantly) All right.
```

They will meet and live happily ever after! I sit up straight, a smile on my face – I like happy endings.

In 2018, at the Tate Modern in London, I stare at *This is Radio Moscow . . .*, a painting by Viktor Pivovarov.

The painting is captioned like a comic, in Cyrillic, which – after six years of Russian instruction in Polish school – I can still read: "The time in Moscow is 19:30. We begin the broadcast of 'Theater at the Microphone.'" The painting depicts a room with a French window open onto a courtyard between apartment buildings on one of those late summer evenings when even at 11:00 p.m. the sky still glows peach-orange. On the table, a tall glass filled with what has to be tea, the national drink Russians have always drunk from a glass held in an ornate wire basket; a teaspoon; an open notebook; a lamp with an orange shade. Only the listener is missing. Across the yard, another tenement building with a single lit widow – what do you want to bet that the people in that room are also listening to the same program? In my gut, in my heart, I remember, no, I feel the nighttime wonder. I am transported back to my dark little room, the green eye of the radio glowing, soft voices murmuring in my head.

But I stopped listening to the radio years ago. After my family emigrated to Canada, four years after I heard the play on the Polish radio, all I could find on the local Ontario stations were music hit parades peppered with advertising. When I could finally listen to all the Western music I wanted, I realized that I missed the satirical sketches, the news programs braided with popular music, the plays, the serialized Polish literature classics read by famous actors. In Toronto, I couldn't find anything that would keep me glued to the speaker, ears pricked, late at night in the dark. Within months, I stopped listening to the radio altogether.

When I saw the painting at the Tate Modern, I realized what I had been missing for almost forty years.

On the radio, the window swishes open. I prick my ears: How did their meeting go?

```
A: Piotr?
P: Anna?
(long pause)
A: Why didn't you come? I waited and waited
   by the clock, felt like such a fool.
P: I waited too. There were so many people:
   couples, families, babies.
A: I looked and looked. There was a tall
   young man, vary handsome. Bright blue
   eyes, a mop of brown hair, broad shoul-
   ders. He must have been waiting for his
   girlfriend; he was clutching a long-
   stemmed rose. Lucky girl.
P: I was there for over an hour, waiting,
   staring at every woman who walked past.
```

> I never saw you. A young beautiful blond
> looked at me for a moment and then turned
> away. And – wouldn't you know, she was
> holding a single red rose.
> A: *(quietly)* How strange.
> *(silence)*
> P: Maybe it wasn't meant to be . . .
> *(birds chirping, sounds of distant traffic
> fade to silence)*

What?! Why would they both lie? When each saw that the other was beautiful, why didn't they want to meet? It made no sense – they should have been happy to see the other unmarred, hale. My mind is whirring – was Anna hoping to be a beautiful bride to an ugly dwarf? She must have known how beautiful she was. She wanted to be a saving angel? I am too young to know about the saviour complex but that's what my child's brain comes up with. What about Piotr – why did he betray Anna, why did he lie? They could have been so happy – two beautiful people, in love.

Miffed at the characters, annoyed at whoever wrote the play, I press the OFF button on my little Grundig. The green light on the dial fades to blackness and I am left bereft and confused, but the feeling of having been a part of something larger than myself lingers to this day.

PS. I have recreated the radio play solely from memory. I have no idea who wrote it, I don't know if it was written expressly for the radio or was adapted from another work – it does have a bit of an O. Henry feeling. I searched the digital archives of the Polish Radio, but to no avail and so it remains untitled and unattributed.

LAUNDRY DNA

"How do you make *krochmal*?" I ask my mother. The silence on the other end betrays her surprise – it's been more than forty years since she last prepared it.

Krochmal is the starch solution used in Poland for stiffening cotton and linen bedsheets and tablecloths. The word in all its forms dates back to the fifteenth century and is a testament to its importance in Polish housekeeping. There are derivative verbs – *nakrochmalić, wykrochmalić*; an adjective; and several related nouns all having to do with starching fabric. In my part of Poland, it even made its way into a colourful imperative – *nie krochmal*, stop telling tall tales. Many were the times that I watched my mother mix and boil the translucent mixture that glug-glugged into the zinc vat in which she would rinse the laundry. I stirred the cooled goop with my finger; when I licked my sticky fingertip, it had no taste.

*

For years, I paid little attention to my laundry as I shoved it into the washing machine. I did remember to separate by colours and managed not to scald my delicates, I handwashed wool sweaters and silk blouses, but that was the extent of my laundry proficiency. My house has been kept clean and orderly and laundered thanks to a devoted cleaning lady and my neat-freak husband. No scrubbing the floors or dusting for me – my house was clean *enough*, clothes mostly wrinkle-free (my husband took his shirts to the cleaners, unwilling to leave them in my erratic care), bathrooms reasonably spotless.

And then the pandemic hit. Rudderless without my work, lacking direction and focus, I lounged on the couch in tattered track pants and a sweater stretched beyond decency. I read voraciously. Baked gluten-free cupcakes for my husband with celiac disease. Joined online book clubs and writing classes, which I attended presentable from the waist up only. Finished edits to my memoir. And yet, by the end of May 2020, I slammed into a wall – bored, miserable, with a writer's block the size of a district. I couldn't concentrate on anything. I suffered through the summer half-dazed and by September I was ready to do something, anything, to pass the tedium. I started to clean the house.

In the basement, I unearthed a load of laundered but badly wrinkled sheets – I had no idea how long they had lain there. I used to send them to be ironed but the woman stopped because of Covid-19. The two-foot-high tangled heap in the corner of my laundry room retained some freshness, but I washed it again. On the internet, I learned that a combination of equal parts laundry detergent, dishwashing powder, Borax and bleach removes the yellowing on bedsheets and pillows. It was meant for soaking in top-loading machines before the actual washing cycle. Somewhat fearful that the concoction would either melt the rubber gaskets of my washing machine or blow up as I prepared it, I mixed it and added it to my front-loading machine, switched on the hottest setting and loaded in the bedsheets. It worked. With a heap of pristinely white, fresh-smelling laundry I had to face the music – ironing. Not a bad way to pass the lockdown, I thought and dug out my Rowenta from the bottom of my closet.

I lugged the ironing board up to the living room from the basement and stood it in front of the TV – might as well watch something while I ironed. The iron hissed steam and smoothed even the most recalcitrant wrinkles and I felt capable, masterful. For eight hours, split into two evenings, I binge-watched *Young Wallander* (one doesn't need a critically acclaimed show for ironing) and ironed four towering stacks of flat sheets, fitted sheets, pillowcases and three dozen linen dinner napkins. I did not use the canned instant starch even though I had three canisters – it just didn't seem right. The iron and the steam sufficed to get the sheets glossy and flat, and I had been known to scorch the instant starch on the soleplate of my old iron, the brown stain never to come off. I even learned from a YouTube video how to fold fitted sheets so that they could be stacked neatly. It was surprisingly easy and quite satisfying.

I felt ridiculously accomplished after I placed my tower of ironed sheets onto the shelves of the linen closet. It, too, got a total overhaul: the shelves washed, tattered pillowcases discarded, old blankets donated to charity. I arranged the sheets by size and weight – cotton separated from flannel, summer from winter; hung a lavender sachet left from my previous burst of domesticity on a nail on the door, tucked another behind the sheets. Also on the shelves: bottles of freshening linen and wool sprays, and a cube of laundry soap, another epiphany. Several rubs with a citrus-smelling,

Borax-containing bar rid my favourite Michael Kors dress of an old stain that repeated applications of Oxi, Shout and the European Vanish could not lift. I almost kissed the little greenish nub with gratitude – the stained dress escaped culling only because I loved it so much.

Somehow, I had become a master laundress.

Which brings us to the phone call to my mother. For my next housekeeping feat, I want to replicate the wonderful feeling of the clean, fresh tautness of childhood bedsheets, which can only be achieved through the magic of *krochmal*.

In Poland, *krochmalling* was easy – my mother dissolved a spoonful of potato starch into a glass of cold water, added the cloudy white solution to a pot of boiling water, stirring until it grew clear, and then poured it all into the final rinse. In Canada, however, trouble begins at the natural food store: I can buy either potato flour (which was the name we used in Poland) or potato starch. By appearance – a brightly white fine powder that I remember explodes into billowy clouds when not handled carefully – I deduce that for *krochmal* I should use potato starch, not the beige grainy potato flour. I buy a pound of it and carry it home, hopeful and a bit apprehensive – will it work in my high-efficiency, front-loading twenty-first-century washing machine?

In the sixties of my early childhood, laundry was an elaborate affair involving a white-enamelled cylindrical washer that stood as tall as five-year-old me, with a roller mangle on top and a red draining hose below. Frania, a diminutive of Frances, was spelled in blue letters on its side. On laundry day, Frania occupied a place of honour in the centre of our tiny bathroom, pulled from its usual place in the corner. My mother filled the machine with hot water from the tap, loaded the laundry and added detergent. A loud hum rose from it when she turned it on. In the kitchen, a thick tin vat sat atop three gas-stove burners and billowed clouds of steam – there, the washed sheets were boiled for at least an hour, on a slow roil, to get them snowy white. No self-respecting Polish woman used chlorine to bleach her whites. The air thick with steam, rivulets of condensation ran down the panes of the kitchen window, puddling on the windowsill. Tendrils of damp hair plastered on my mother's forehead as, red-faced, she lifted the steaming sheets from the vat with long wooden tongs and dropped them into the *krochmal* filling a metal basin the size of a life preserver. Then she fed the sheets back through the wringer twice until they were ready for hanging. When she finally drained the washing machine through its hose into the bathtub, the water gurgling down the drain, bubbles dotting its surface, a mound of moist sheets sat piled in the basin.

The iron skeleton key screeched as it let us into the communal drying room in the basement of our five-story building – there wasn't enough room in our apartment to dry the sheets, the duvet covers, the pillowcases. My mother would sign up for the key with the janitor and plan her laundry day around its availability. A light, crisp smell of wetness permeated its grey cement floor and whitewashed walls – traces of the thousands of bedsheets that had dried in the small space since the beginning of the building. As she struggled to hang the sodden sheets, I ran between them, running my hands along the wet, and she scolded that I was dirtying them. I loved the damp feel of wet cotton, its smell tickling my nostrils, the wetness cloying my face as I pushed it into them. With the moisture evaporating from the laundry, the room smelled like wet sidewalks after a spring shower.

Two days later, we descended to the basement again to lift the sheets off the lines, stuff them into a large wicker basket and carry them up to our fourth-floor apartment. The *krochmal* tightened the fabric and the sheets needed to be stretched out to their original size. My father would be corralled for this task when he returned home from work in the evening – it was a two-person job and my hands were too small to grab the corners and too weak to pull hard enough to straighten the fibres and the hems. Facing each other a sheet-length apart, corners clutched in their hands, my mother and father pulled and pulled in rapid, swishy strokes, alternating hands, the poof, poof, poof sound making me giggle. Then, switching hands, they folded the sheet lengthwise and, as if in a dance, approached each other, their hands meeting in the air and my mother plucking the corners from between my father's fingers and lifting them high. Without missing a beat, my father bent down to catch the bottom fold to raise it off the floor. They repeated the manoeuvre one more time to fold the sheet again and then my mother would finish by herself, halving the sheet twice more and dropping it into the waiting basket. Repeat for each flat sheet and each duvet cover, every pillowcase. Sometimes, at the beginning, when the sheet resembled a hammock hung between their hands, I would be allowed to climb in and, laughing, they would toss me up and down, my belly squirming, my limbs flailing. Released from the sheet, I sat on the bottom of the wicker laundry basket, which was large enough for me to fit in, the folded sheets piling on me until the basket was filled with waves of fresh-smelling fabric. It was like being buried in warm snow.

The next day, my mother and I took the folded laundry to be steam-ironed. At the local *magiel,* an industrial-size rolling steam iron erased the wrinkles out of the fabrics, leaving them shiny and silky smooth, a feeling I remember from when I burrowed into their crisp, fragrant embrace for the night.

*

The bedsheets of my childhood snap into my mind as I stand by my ironing board. Others are reliving memories of their childhoods by baking bread and assembling jigsaw puzzles, but I reminisce about *krochmal*, the dripping windowpanes, the rolling cylinder at the *magiel*. The clean scent of boiling detergent and of the basement drying room. And one evening, as I press my iron over a stubborn wrinkle, I begin to think that it might be more than a simple childhood memory.

Before World War II, my maternal grandmother worked in a pension in Niemirów-Zdrój, a spa town nestled in the fir-and-beech forests of southeastern Poland. In a photo I have of her, she stands on a wooden ramp with a group of women in white smocks, all with white kerchiefs tied behind their heads. They seem to be caught mid-jostle, these young women who look like laundresses; all are laughing. I imagine them boiling the bedsheets they had stripped from the beds that morning and the towels they picked up in the bathrooms and the change rooms on the tiny lakefront beach. The tin vats filled with soapy water are huge, fit to stew a calf whole. I imagine the stifling hot laundry room, the women just grey ghosts flitting behind the billows of steam. I hear the scrunch, scrunch, scrunch of the sheets on enormous washboards, an unyielding rhythm of their arms, muscles thickened from the scrubbing, the redness of their hands a bloody contrast to the white sheets. The lye and detergent split their nails and dry their skin, nothing helps though they have spread Vaseline and lanolin cream on their hands; some have even tried butter smuggled out from the dining hall. Later, the sun beating on the sheets strung on the lines in the garden will finish the job of getting them brilliantly white, the sheets squeaking from the *krochmal* as the women hang them and clip them with wooden clothespins. The sheets flap on the lines stretched between tree trunks and the women will take them down before the evening dew, after they finish the dinner service – the summer evenings are long and warm at this latitude. They iron the sheets the next day with heavy slug-irons. The oblong iron "soul" – the Polish term for the insert – would be heated in the stove and, red-hot, yanked out with metal tongs and slipped into the iron's box. One had to be very quick stroking the fabric so as not to burn it.

I'll never know for certain, but I think that the bedsheets smelled just as fresh as a river after a spring shower when the heads of the pensioners settled on them at night.

The next time I wash my bed linens – I choose a sunny Saturday in late September – I carefully measure out the potato starch so as to avoid the powder explosion and mix it into a glass of water just as I remember my mother doing it. The water turns thick and gloopy, but remains translucent. So far, so good. I pour it into the fabric softener dispenser. I select an extra rinse, start the hot wash cycle and cross my fingers. Two hours later, the laundry comes out of the drum squeaking slightly – I remember that sound! – as I pile it into my brand-new wicker basket bought online from Etsy, a pandemic purchase, and carry it to dry on the deck.

When I lift the sheets and pillowcases off the drying rack at the end of the day, they are stiffly holding their folds just like those sheets in the basement drying room. They crunch under my hands. It worked!

My grandmother died of tuberculosis three years after the war ended, when my mother was eight. Sulphonamides did not arrive in Poland soon enough to save her life. But I can smell her in the freshness of my clean bedsheets, see her smile hover in the steam hissing from my iron, feel the touch of her fingertips when my cheek lands on the smooth, crisp, starched pillowcase at the end of the day. These sensations seem to be embedded deep, deep inside my marrow, transferred through generations, and all I can do is hope that my sons will remember those sweet-smelling sheets I have ironed for them.

A KNITTING CLASS

I never met my grandmother: she died when my mother was eight years old. All I know about her is my mother's stories and a handful of photos that survived the war.

This photograph is dated March 30, 1935; a yellowed cardboard rectangle softened with age, *A memento from the course* and the date written on the back in my grandmother's neat handwriting. It was taken in Niemirów-Zdrój, a hamlet fifteen kilometres east of the present Poland-Ukraine border; at that time, it belonged to Poland. The village, nestled in tall pine forests on the sandy shores of the Bronka Creek known by its sulphurous smell, was a mineral waters spa popular with the bourgeoisie from Lwów, the province's capital. My grandmother was seventeen and a half at the time and one of fourteen girls in a sewing and textile crafts course: they learned knitting, crocheting and tailoring basics.

In the photo, my grandmother is sitting front row, left of centre. She is the one with the broad and high forehead, and the strong and determined chin; she wears a dark sweater and grips a knitting needle in her right hand. I am struck by how much I resemble her, all the way down to the step in the middle of her hairline above her forehead. When I show the photo to my husband and ask him to find my grandmother thinking that he would see the resemblance, he says that he doesn't know which one is my grandmother "but what the hell are you doing sitting there?" Incredulity rings in his voice.

The girls sitting around the rough-hewn table piled with fabric samples and balls of yarn, with a treadle sewing machine at the front, its needle piercing a swatch of fabric, could be Polish, Jewish, Lutheran or Ukrainian. Before World War II, nations and religions intermingled freely in the small town. Some of these young women probably didn't survive the vagaries of war – first the Soviet, then the German occupation of their little village, the Nazi-supported terror rained on the Polish civilians by Ukrainian nationalist bands, the resettlements during and after the war. After the Soviets occupied Niemirów, the Polish girls might have ended up with their families on cattle-car transports to Kazakhstan, and later to Iran, Africa or Australia; when Germans came a year and a half later, the Jewish ones were likely rounded up into the nearby Rawa Ruska ghetto and sent to the death camp in Bełżec.

Two years after the photo was taken my grandmother married. When Germany and the Soviet Union invaded Poland from opposite directions in September 1939, she was already a mother to a one-year-old daughter, Lusia. Her second daughter – Krysia, my mother – was born in the town's church cellar during the bombing raid that began Hitler's invasion of the Soviet Union in June 1941. The Nazis had reneged on their non-aggression pact with the Soviets. Soon after, her husband was conscripted into the Red Army. Poorly trained and shoddily equipped, my grandfather was taken prisoner by the advancing *Wehrmacht* and survived the war in a Bessarabia prisoner of war camp until it was liberated in 1944.

Under Nazi occupation, in the fall of 1942, my grandmother buried her father-in-law who was too old and too frail to last through another winter of near-starvation, and in the winter of 1943 she buried Lusia who died from meningitis. In July 1943, when Poles were ordered at the fear of death by Ukrainian nationalists to leave, she escaped with Krysia into the General Government part of occupied Poland and found shelter in Leżajsk, a town beyond the border on the river San. There, in a one-story house emptied of Jewish lives, she eked out an existence for three years. She caught pigeons and cooked broth for her daughter in the spring, she dug potatoes from the frozen fields in the fall. When the Soviet front stopped on the river, she sewed the cigarettes her landlady got from German soldiers she befriended into the lining of her coat and punted a small boat across the fast-moving river to sell them to the Soviets. She didn't know how to swim. She was in danger not only from the Nazi police and the Soviets, but also from the Polish underground army who would have executed her for collaboration had she been caught.

My grandfather, freed from the camp by the advancing Red Army, made his way to Leżajsk in late 1944. As the newly carved borders in Eastern Europe swung west, Poles were resettled in another cattle-car exodus, this one westward and mandated by the Yalta accords between Roosevelt, Churchill and Stalin. Thus, in 1946, my grandmother's little family found itself almost four hundred kilometres west, in Gliwice, a city in southwestern Poland where they were assigned yet another empty flat, this one abandoned by the Germans fleeing the Red Army. Here, in 1947, another daughter, Danusia, was born and placed in an orphanage because of the galloping consumption my grandmother had developed by that time. In early 1949, my grandmother was admitted to a tuberculous sanatorium in Bystra on the high foothills of Klimczok, a peak in the Silesian Beskids mountains eighty kilometres south of where she lived. She never saw either of her daughters again. In the fall of 1949, when her condition became terminal, she was transferred to an isolation ward in a hospital in the nearby city of Biała. She died on November 30, 1949, far away from her birth village and from her transplanted family, quarantined on an infectious diseases ward, alone.

But the serious girl in the photo doesn't know any of that. She stiffly holds her knitting needles and her swatch for the photo, her righthand knuckles white with tension and a thread tightly wound around her left index finger. Right after the click of the camera I imagine that she breaks into a toothy smile and stretches her stiffened shoulders and back, shakes out her hand. Her friends, unaware of what history will bring their way, chatter and joke, the teachers shush them, but they, too, are relieved that the posing is over. Life returns to normal, even if just for a moment.

THERE BUT FOR THE GRACE OF GOD . . .

On the computer monitor, the fuzzy black-and-white ultrasound images coalesce into an incomprehensible grey jumble of a bifurcated skull, a tongue lolling out of a gaping mouth and stunted arms. The normal-looking legs and feet dangle from the lumpy mass floating in the amniotic fluid, clearly not belonging. The pregnancy is eighteen weeks along.

I hear the radiologist mumble something about losing his lunch. My eyes pop open and I'm not quite sure where to look in the darkened solitude of the diagnostic imaging room. I realize that the clatter in my chest is my heartbeat.

Just like mine, his mind is trying to process what we are seeing: fetal malformations that defy the normal body plan, incompatible with life. His comment isn't about gastrointestinal discomfort but an expression of the distress we are both feeling at seeing something so abnormal. I glance around and to my relief, we are alone.

Before the advent of prenatal ultrasound, this fetus would have lengthened for the next four and a half months, grown to full size only to be delivered stillborn to the dismay of the midwife and the parents. For millennia, such babies have been seen as a warning of impending doom, an allegory for the sins of the populace or proof of maternal infidelity. The birth of this baby would have been as amazing as the birth of a monster in Ravenna in 1512, which reverberated throughout Renaissance Europe. A monster, yes, for that had been the technical term used for centuries by physicians, philosophers and in modern clinical medicine as late as the mid-1980s.

A week later, as I examine the stillborn fetus, I diagnose amniotic band syndrome: tough, fibrous ribbons of tissue have detached from the lining of the uterus and, floating in the amniotic fluid, wrapped around the developing limb buds and head. Just like a wire looped around a tree branch cuts into the bark as the tree grows, these bands cause amputations of limbs and digits. If swallowed, they cause cleft lip and palate. The only good news is that amniotic bands are very unlikely to happen again and I will be able to reassure the parents that they have a very good chance of having a healthy baby in the future.

I am a clinical geneticist. In my practice, I see malformed fetuses and newborns all the time. I consult on newborns born with unexpected, always unsettling, often shocking features. And even though I have seen hundreds of such nature's creations, I still haven't seen it all.

Since childhood, I have been fascinated by these errors of human formation. Does that make *me* a monster?

I had my fill of monsters when my father read the Greek myths to me as bedtime stories. One-eyed Cyclops, lion-bodied sphinxes, snakey hydras and winged harpies ignited my imagination. I relegated them to the land of fairy tales, unaware that such creatures had been seen by the ancients in real life before they became the stuff of myths and legends. It was only in medical school when I was asked to explain why the Cyclops was an embryologic impossibility that I realized the true origin of many fabled creatures. But those were not as intriguing as the drawings and photos in the medical handbook I snuck peeks at whenever I could. Among the colour plates of rashes and black-and-white drawings of skeletal defects, I discovered what looked like remnants of photographs in the page gutters. They appeared to have been cut out with manicure scissors. When I asked my mother what had been there, she said they were photos of abnormal babies and that they were so disturbing she couldn't look at them. A few years later, when I found an unmutilated copy of the same handbook on a neighbour's bookshelf, I searched for those forbidden images. A baby with a massive hydrocephalic skull lying in a crib, her eyes half sunk under the lower lids, sunsetting. Three naked boys with coarse-looking faces, each with a huge goitre, their tongues hanging out lifelessly between thickened lips. I pored over the book, holding my breath until I could no longer, my belly pounding with my pulse, and then shoved the book back where it belonged in the bookcase.

Abnormal-looking newborns have captivated and terrified the witnesses of their birth for as long as the human race has existed. Such births were viewed either as portents of disaster or incarnations of evil, an attitude that would not change for millennia. At the same time, surviving people with anomalies were made members of aristocratic courts such as Antonietta Gonsalvus, the subject of Lavinia Fontana's 1595 painting *Girl Covered with Hair,* or the people with achondroplasia depicted in Goya's paintings of the eighteenth-century Spanish royal court.

Collecting biological specimens that differed from the accepted norm became a popular pastime for the aristocracy during the seventeenth and eighteenth centuries and gave rise to the cabinets of curiosities. The Holy Roman Emperor Francis II amassed a sizable private collection of curiosities in Vienna, currently housed in the Narrenturm, an old hospital. Others exist in Bologna, in London, in Saint Petersburg, all stemming from such collections. It wasn't until Francis Bacon recognized that the study of errors of human development revealed the laws of nature governing normal processes that a more rational approach prevailed, laying the foundations for the study of embryology and teratology.

I visited the Narrenturm on a bright April morning in 2016. The Collection of Anatomical Pathology is housed in a round, four-story tower, the old psychiatric ward of the Vienna General Hospital, separated from the main building by an expanse of lawn. It was a fitting, if gruesome, setting – thick, whitewashed brick walls separated tiny cages of rooms on the outer wall, a circular corridor surrounded an inner courtyard: Was that where the less affected inmates had been allowed to take air? I had expected the place to reek of formaldehyde, like the pathology departments in all the hospitals where I had worked, but the building was odourless, sterile. The specimens of birth defects floated in two-hundred-year-old glass jars filled with murky fluid: babies with skulls expanded by hydrocephalus; a small mermaid with its legs fused, the joined feet splayed into a fishtail; a baby with its intestines floating outside its abdomen, its little fingers interlaced as if in prayer, put in that position by some well-meaning – or was it morbid? – mortician, and slumped forward, its nose flattened against the glass of the jar, its knees bent back so that it was kneeling in its glass cage. In medical school, I had seen such exhibits in our anatomy department, used them as mere learning tools. But all these babies had had parents and siblings who outlived them but then, in turn, they also died, and now, many generations later, only those that did not survive were preserved for posterity. Their mothers most likely never saw them, whisked away as they were for their protection and to be studied and embalmed, never held, never cherished. Rattled, I skipped the guided tour of the entire collection that was to start half an hour later and sat on the grassy expanse for a long, long while, thinking about the abnormal babies that I had seen wrapped lovingly in blankets, crocheted hats perched on their heads, and given to the grieving mothers and fathers to hold. All cherished and all mourned.

But it is not the history of monster-watching that fascinates me. It is the fascination itself – why are we drawn to monsters? Why is it so hard to avert our eyes from a person with an unrepaired cleft palate? Why are toes fused into flippers more fascinating than normal feet? Why do people visit the Narrenturm? Why did I?

I once saw a deaf-mute signing for alms in the archway of an old building on the main street of my hometown. My five- or six-year-old eyes fixed on his gaping mouth, my ears strained to hear his guttural utterances until my mother yanked my hand to make me move. Fear and – yes – curiosity coursed through me. He was thin to the point of being emaciated, hair closely shorn under his peaked grey cap, the fingers of his right hand held together at fingertips pointing toward his open mouth in an instantly, instinctively recognized sign for "hungry." On the way home, I plotted how my grandfather who rented out rooms to university students could take him as a lodger so that the man could get a job at the State Disableds' [*sic*] Cooperative, which happened to be located just down his street. The poor man could live and work in dignity, and not have to beg on the street anymore – such a simple solution. Why hadn't anybody thought of it, I wondered as I lay awake into the night.

In our tenement building, there was a teenage boy who had wobbly, unruly arms and legs that flailed about when he tried to do anything. He pushed around our yard on a battered beige stroller that he propelled with his right – the stronger – leg. His face contorted into a grimace when he produced his garbled sounds, but we understood him. When he laughed, his face split into a freakishly wide grin, teeth and purple gums showing. He played hide-and-seek and catch with us, wheeled around with a posse of boys in the surrounding forests. His name was Heniek, but we called him "Heine-Medina," the eponymic used for childhood polio of which there were many survivors after the epidemics in the fifties. In medical school, many years later, I realized that he probably had choreoathetotic cerebral palsy that stemmed from untreated newborn jaundice, not polio. Heniek didn't engender those feelings of fascination or fear. Different, he was still one of us.

It was the ones born differently that fascinated me and caused a frisson of fear. Was it a fear of being like them? What made me lucky? What made me different? And why? My body was whole, my brain worked, I could speak and read and behave. If I were different, I'd be stared at, fingers pointing, be the target of sidelong, embarrassed glances. The societal norm dictated that you didn't stare openly, but still, you would see: flailing or missing limbs, distorted faces.

Despite that disquiet, I kept looking. Curiosity overcame the fear. I am told that I first announced that I wanted to be a doctor when I was five years old. In school, my favourite subject was human biology. I wanted to know how the body was put together and how it worked, how cells were arranged into tissues and those again into organs, and how it all functioned. I couldn't wait to study anatomy, biochemistry, histology, physiology. I wasn't interested in diseases or their treatment – it was all about the marvellous wonder that is the human body and the magical way it worked.

*

When I arrived in Toronto as a sixteen-year-old, in the early eight-ies, I was bewildered by the ads on the buses and on the subway for fundraisers for diseases I had never heard about. Names like cystic fibrosis, Duchenne muscular dystrophy, Down syndrome. I had a vague knowledge of *mongoły* – the term used in Poland for individ-uals with Down syndrome, but what were the other diseases? Were they really so common here that they required the public's help? In Poland, any form of bodily or mental difference was considered shameful. Such individuals, adults and children both, were kept indoors, away from strangers and relatives, they did not participate in telethons or sit for photos that would appear on subway posters. Squirming on the orange plastic seat on the Queen Street West streetcar, I wondered if Canadian society was sicker than Polish. At home, I looked up "cystic fibrosis" in the English–Polish dic-tionary: *mukowiscydoza.* I had never come across that disease in all of my readings. The next time I saw my family doctor, I asked her about it; she also had emigrated from Poland. She enlightened me about the burden of genetic disease in Poland. "You have no idea how many sick children there are," she said. Medical care for them was dismal, she added.

Like for the two girls I remember from my later chidlhood. They slouched around our apartment complex: hair uncombed, tattered summer dresses with yellow-stained panties peeking underneath, their faces smudged with dirt, hands goopy with unknown sub-stances. They were unusual because they were out in the open, not hidden away from prying eyes. They tagged behind a stout woman who might have been their grandmother. One emitted animal-like moans and darted toward passersby, grabbing at their clothes; the other uttered monosyllables. I assumed they were sisters. They frequented our playground in the summers and disappeared into some unknown place during the cold months. I wondered about their parents. I wondered what they ate and where they slept. The woman yelled at men who walked by on their way from work and at the gossiping women seated on the benches in front of the buildings, but she never begged. None of them appeared hungry, but the girls sometimes snatched toys from the smaller children playing in the sandbox.

I never learned their names.

At university, although I was a pre-med hopeful, I hadn't enrolled in the genetics course because the teacher was known to be a very hard marker, so it wasn't until a classmate brought his textbook to our botany class that I saw what it contained: photographs of babies and children with genetic disorders. My dormant fascination stirred and woke up, this time for good. As I flipped through the pages, another classmate caught a glimpse of a photograph of an unusual-looking hand and commented on how pointy the fingers were. "Who cares if they're pointy?" the owner of the textbook asked in a high-pitched voice. "They have no nails!" He sounded shaken.

He was wrong – there were nails on the fingers but the photo was taken from the palm side of the hand. I figured it out two years later when I used the same book for my medical school genetics course. The photograph was meant to showcase the abnormal loops and whorls on the patient's finger pads, not the top aspect of the hand where the nails grew.

The embryology course in first-year medical school convinced me that there was a higher power. For a fully formed and functioning body to develop from a single cell, somebody, something must be watching over the myriad processes that could go awry at any picosecond. That the process of putting together a whole human being went wrong only a minuscule fraction of times was proof enough for me. The embryology textbook also had examples of things going wrong, in the Baconian spirit of illustrating the normal by studying the abnormal. There were photos of spina bifida and anencephaly, of cleft lips and club feet.

The strange thrill at seeing a malformed face or body, the "can't look away" feeling happens to all of us. For a while, in medical school and early in my residency, I wanted more and more, like an addict jonesing for a hit with diminishing returns, until one day the photos didn't bother me anymore. There were times when this fascination bordered on obsession. During my genetics fellowship, especially, I worried that there was something wrong with me. I would get excited when I described the differences in the patient's physical features and, once recognized, those features clinched the diagnosis. I would expound on them and my mind delighted at their coalescence into a known clinical entity that we understood and could explain. I was only interested in making the diagnosis. I wanted to learn, learn, learn. The fascinated child was thrilled by the findings, by the new knowledge; it was the excitement of a researcher on the cusp of discovery, facing the challenge of the unknown. I knew that these children were human beings, and I felt sorry for them and for their parents, but I had a job to do. To examine their little bodies for clues; to evaluate every detail, every photo, X-ray and laboratory result; to scour the published medical literature to make a diagnosis and to explain it to the family.

A question at the end of my training: When does a severely malformed fetus cease to be human? An instant answer: Never.

"Can you show us what you mean?" asks the father with an insistence not dampened by the beeps of the NICU monitors and the whooshes of the respirators. He has his arm around his wife's shoulder as if steeling her for the onslaught of new reality that the birth of their third child has brought.

I have just told them that I thought their newborn daughter had trisomy 18. "It's because of the features that she has," I said.

He is a geography professor at a university in a neighbouring town, I later learn, and is rational and scientifically minded. He wants to see what I am seeing. He comes closer and puts his hand on top of the plastic box of the incubator.

I pop open the portholes into the Isolette in which their daughter lies, an aquarium-like box where the temperature and humidity can be controlled for the benefit of the sickest newborns. I gently pull off the tiny pink crocheted hat and lift her head.

"See how the back of her head protrudes?" I ask. They nod. "Her face is tiny: it's almost as if all of her features were in the middle of her face."

"And here?" I run my finger along the baby's sternum. "The breast bone is very short and ends very high in the chest." I point my finger to its end. I move my hand down the baby's tiny body and lift her left foot. "See how the sole is rounded out? We call it rocker-bottom foot. It's characteristic of trisomy 18." I reach for her left hand, the one free of the intravenous line. "I can't straighten her fingers. See how the pinky and the index finger overlap the middle two fingers? That's probably the most characteristic finding."

I tuck the hand in, click the portholes shut and face the parents. "I am quite certain that Emily has trisomy 18, but we need to wait for the results to be completely sure."

The father thanks me.

At times, especially at the beginning of my career, cracks appeared in my professionalism. I am not made of granite; I'm more like a soft, crumbly limestone and it takes strength to hold it together. There were times when I had difficulty bridling my enthusiasm, my eagerness at observing new findings or at learning new conditions. There were tears at the lethal diagnoses and shock at the gruesome malformations. There was anger at the unfairness of the biological processes that left the babies so affected. Professionalism at the bedside means holding back and being stoic. It protects the patients and doctors from our unbridled emotions, but feelings are important: we need them for empathy.

In the first year of medical school, I had to look up "empathy" since I had never come across the word before. It might have been my limited English, at that time only four years after I arrived in Canada, but I suspect that I wasn't alone in my class, uncertain about its exact meaning. Our clinical skills handouts defined "empathy" as "the ability to understand and share the feelings of another," but that definition shed little light for me. How did it differ from "sympathy," a word which translated to Polish as "co-feeling" and which I understood? The prefix, "em-," confused me. In the interviewing skills course, to hone our empathy we were taught how to listen attentively – to use verbal and non-verbal cues to encourage and further the conversation. We were taught the difference between open-ended and close-ended questions. At the time, I thought it was contrived and phony, a fake professionalism, like the instance several years later when a teacher admitted to "putting on the waterworks" when giving bad news to parents. Now that was monstrous. From the beginning of my training, I believed that if I felt empathy for the patients (whatever it meant), I would not need those verbal acrobatics, I would not have to pretend to experience feelings I wasn't feeling. Compassion should show through words and actions. Of course, during clinical exams in medical school, I went along with those suggestions to pass the course, but in clinics, with real patients, I seldom used them. Instead, I listened. I talked to them. Listened some more.

*

In the autopsy suite again, I examine another baby.

The rosy fetus, its translucent skin glistening under the harsh fluorescent lights of the pathology department, appears perfectly formed from the pelvis up. It is a boy. His belly and chest are proportionate and his arms end in perfectly formed fingers, each with its own pearl of a nail. The head – not much larger than a hen's egg – is nice and round. The forehead bears beautifully arched eyebrows over the fused eyelids. A lovely nose and red lips finish the picture that would be a paragon of health and perfection if it wasn't for what is missing: the baby's legs are replaced with tiny nubbins attached to the pelvis on either side of the genitalia, both not much larger than the baby's scrotum. The left-sided appendage ends in a single toe with a fully formed nail, as if nature tried really, really hard to create a normal foot.

I stroke the soft head as I look for clues to what went so terribly wrong. No signs of a syndrome, no amniotic bands. I hope that I will find a spelling mistake in his DNA that will explain what had happened, but I realize that I may never be able to give his parents the answer as to what happened to this tiny mite.

It turns out that I am right – all the available twenty-first-century molecular genetic tests yield no diagnosis. I have no explanation for his defects. Diagnostic failure and sadness accompany me to my consultation with his parents a few weeks later. All I can do is listen and explain. Sit in silence with their tears, if that's what is needed.

I know how to do that.

THE ART OF LOVING

The slap on my left cheek echoed in the darkened vestibule of our apartment building, my father's indistinct outline looming in the shadows behind my mother.

"Where have you been?" she hissed.

Where? Behind the neighbouring apartment tower, making out with a boy from my school, my first non-spin-the-bottle kiss. Marek and I had snuck out behind the eleven-story concrete building dotted with tiny windows and half-balconies. Snuggled into a corner where walls met, we explored this kissing business. I was thirteen and Marek a few months older. He was not my boyfriend, we were not going out, he wasn't even that cute but something about his demeanour enticed me. He was a bit of a player, escorting a different girl from school every other day and parading around the neighbourhood with an arm slung around their shoulders or clutching their waists or even! cupping their nascent breasts. I was flattered by his attention. I was as good as Iwona, our school's "it" girl that all the boys wanted to take out. Tall, willowy, her narrow ass tightly snuggled into the Levi's her father had brought from London, England. She played tennis and was the school track-and-field star. I was none of those things but suddenly Marek was paying attention to me. His first tentative kisses quickly became an excited full-on exploration of my mouth – his tongue inside my lower lip, climbing onto my tongue, my mouth opening wider and wider. My back pushed against the coarse cement that scratched through my thin summer dress, I yielded, a tingling sensation in my lower belly and, soon, wetness where my legs met. We lingered in the falling dusk, his hands on my ass, our breathing fast. When he pulled back and asked if his moustache – a never-shaven fuzz under his nose – excited me I failed to hide my smile, it was such a needy, naive question. The spell was broken. He didn't even walk me back to my apartment building half a block away, at the entrance to which my mother, my bemused father in tow, had lain in wait.

"I've been looking for you for hours!"

It hadn't been that long, maybe forty minutes?

"I should have never given you that book, you slut," my mother spat out.

My eyes popped open in utter shock.

She stomped to our apartment and I followed, cheek stinging sorely but not as much as my self-respect.

The book was *The Art of Loving* by Michalina Wisłocka. It was the first ever sex-education manual published in Communist Poland. Its bootleg photocopies circulated clandestinely between neighbours and friends both before and after its official launch in 1978: a copy of the typeset was leaked from the publishing house before its official print date.

My mother had thrust the borrowed paperback at me earlier

that summer. "I might as well give it to you, you'll ferret it out any-way," she had said, sounding angry. Was she on the vanguard of the sexual revolution or was she simply relinquishing responsibility for my sexual education? Remembering her halting, crimson-cheeked "Do you know that babies come out your private parts?" when I was five, I suspect it was the latter.

I devoured the flimsy paperback in a matter of days.

I had no idea what a tortuous road the manuscript had travelled before it landed in my eager little hands. Even before she wrote it, Wisłocka, an obstetrician-gynecologist, had been too outspoken, too modern, too independent for Communist Poland. In retali-ation, the censors stalled the book's publication for several years. Polish sexology experts, all male and middle-aged university pro-fessors and her competitors in private practice, had decried Wisłoc-ka to be a deviant pervert influenced by the West and warped by its decadence. Not to mention a witch and a nymphomaniac. Her previous failed marriage provided ample proof that she was a loose woman not to be allowed to contaminate the moral Polish society. The censors demanded that she not write about pre- or extramarital sex – a hypocrisy since most of the high-level Party members were unfaithful and lecherous. When her critics finally relented, it was with a suggestion that the manual be castrated of the chapter on female orgasm; the state-hired scientific reviewers asserted that "a female orgasm is not required for procreation." Wisłocka vehe-mently vetoed their demands and the publication stalled again.

In her inimitable style, Wisłocka cajoled and threatened, even attempted blackmail. From her Warsaw clinical practice, she had information to spare on the intimate lives of many higher-ups. Par-ty officials who were infertile, who beat their wives, who couldn't get it up. In the end, it was her patients – the wives of highly placed

Party apparatchiks – who convinced their husbands to allow the publication to proceed in its entirety in 1978.

Unrelenting, the censors demanded an "elegant and modest" cover. The artist delivered on both points. On a pastel background graded from fuchsia to pinkish white, a line drawing of a groom's and a bride's heads, hers in a virginal veil topped with a myrtle wreath knotted with flowers, white of course. The heads, drawn from the back, lean toward each other tenderly – a picture-perfect image of morality and decorum. No suggestive poses, no kisses, lest the readers get the wrong message. The book's official subtitle hammered the point home: *A Practical Guide to Married Bliss*, leaving no doubts about the marital status required for the enjoyment of the book's contents. Good Communist marriages were pious and productive of many children for the glory of the Party and the nation. But the guiding principle of *The Art of Loving* had always been love as a foundation for great sex, its *sine qua non*. It was not just an afterthought written to appease the Communist censors – love is woven throughout the work. Love and respect, the latter infrequently accorded Polish women.

According to the front matter, the book's first printing was ten thousand copies, but in fact, it was ten times higher: the censors feared making the enormous number public. They did not want to be seen as disseminating smut. In the end, seven million copies were sold in Poland alone and it was translated into most of the languages of the Eastern bloc.

By the time my mother gave me *The Art of Loving*, I knew more about the biology of procreation than most Polish teenagers ever would. I had been poring over the *Little Encyclopedia of Health* since before I could read: first examining the photographs and diagrams and colour plates, then slowly putting syllables together to decipher the text, including the chapter on the reproductive system. In grade eight, we finally had a course on human biology. During the first class in September, Mrs. Antończyk, our biology teacher, loped into our fourth-period class and asked for a copy of the textbook. Tomek, my deskmate, handed her his.

She lay the hardcover book on its spine on her desk and held it vertical between her palms. When she let go, the covers splayed like a bird's wings and the pages fluttered open about two-thirds down the volume.

Mrs. Antończyk chuckled. "Whenever I pick up any of the grade eight biology textbooks at the beginning of the school year, they always fall open to the chapter on human reproduction," she said. We all laughed and Tomek's cheeks reddened.

"I'd rather you learn about it from the book than on the street," the teacher said. She scanned the boys' eager faces dotted with acne and sporting wispy shadows under their noses and glanced at the girls curving forward over their desks to hide the breasts that had blossomed over the summer. She nodded. "We'll get to it soon enough but first we need to learn how cells work."

But it wouldn't be until I read *The Art of Loving* later that year that I would see a drawing of an erect penis – stuck out, perpendicular to the torso. In a section through the body's midline, the drawing depicted the four stages of erection: at rest, minimally stimulated, fully erect and post-coital. How did men manage those things sticking out like that? Until then, I had only seen penises at rest, so to speak, saggy worms pointing modestly to the ground in my parents' art albums: ancient Greek statues and Michelangelo's *David*, the numerous nudes of Baroque paintings. I had thought that those floppy little worms cuddled themselves into the folds of the vulva and delivered their payload thus; now I was shocked by the pointed savagery of them, their bloody-minded directionality, lifted away from the softness of the underbelly. Erect, they looked sentient. They looked – dangerous.

*

As a tween about to embark on puberty, I had a very simple plan with respect to sex – I was going to do it with the man I loved and would later marry. My mother needn't have worried. I wasn't going to wait till my wedding night, no, that would be old-fashioned, but the man to whom I would give my virginity would be the man I loved. I imagined it would be with my teenage crush – tall, blond, blue-eyed Cezary who lived three stories above me and went to the same primary school as I did. I had pined for him since the moment I saw him run up the steps of the building we moved into when I was ten. He was one of the boys who took Iwona out. They often played tennis together, the pastime of the very well-off Communists. And even though we ended up in the same high school and often walked to school together, I never rated.

At the end of August 1980, after I finished grade ten, my family left Poland. In Austria, where we stayed for six months in the limbo awaiting approval by Canadian immigration, my sexual education continued. The seedy inn in Sankt Georgen im Attergau in the foothills of the Austrian Alps where immigrant families were billeted was a melting pot of refugees of all ilk from Eastern bloc countries. I met Ewa there. She was seventeen but looked older and carried her sizable breasts with enviable confidence, mine being much less noticeable. She sought me out and soon was regaling me with her wisdom. "It really hurts down there after sex," she announced one evening when we sat in the deserted dining room playing gin rummy. "So bad, you can't wash. You know you have to wash real well afterwards, every time?" She stared at me sternly. I didn't remember anything of the kind from reading Wisłocka, but Ewa's conviction was such that I didn't contradict her.

It was in that small subalpine village that I saw my first real penis when I got a part-time cleaning job in a pension across the street. One snowy morning, as I poured the soapy water from scrubbing the floors into the sink in the basement kitchen, a wooden door opened at the end of the hallway and, in a billow of steam, a naked man stepped out. I hadn't realized that there was a sauna there. The little wilted member trembled as the man trudged across my line of sight, his eyes scrunched myopically. He never noticed me.

As the winter of our enforced waiting continued, Ewa and I hiked the snow-covered hills surrounding the little town. Some afternoons, one or more of the young Polish men accompanied us. Thanks to Ewa's innuendo-laced talk, the conversation often slid into matters of sex. About three kilometres away, the E60 crossed the countryside on its way from Vienna to Salzburg and we often stood on the bridge that spanned the four-lane highway, waving at the passing cars. One glorious, frosty Sunday afternoon, the sun reflecting diamonds on the crusted snow, a twenty-four-year-old law student from Kraków scoffed when I told him about my virginal marital plans.

"Are you sure that he'll appreciate it?" he asked.

Suddenly, I wasn't. That's what happened in all the novels I had read and that was what *The Art of Loving* had said. I should've answered that the man I chose would, of course, but my cheeks tingled as blood rushed to them and, mortified, I clammed up.

He sauntered back to the inn with a satisfied smirk and I avoided him until he immigrated to Australia a month later.

Wisłocka had lived my dream – she married her first love. In the 1930s, Michalina Braun, still in high school, met Staszek Wisłocki, a biology university student. He was handsome, he was athletic, he was brash and superconfident. He insisted on having sex as soon as they started dating and deflowered her so brutally that Wisłocka said it left her afraid of sex for years afterwards. Despite that trauma, she married him – she was in love. They moved to Warsaw and lived through the German occupation there. Wisłocka obtained her medical degree in 1952 even though Staszek continually asserted that his – middling – academic career in biology trumped hers and that two years of medical studies sufficed to become his assistant, which was all that she was fit for anyway. Undeterred, she finished her specialist training in obstetrics and gynecology in 1959 and continued her research in cytology. Her first publication, titled *Technique of Pregnancy Prevention*, concerned a taboo subject in both Catholic and Communist Poland where birth control methods consisted of withdrawal, condoms and, only much later, spermicidal vaginal tablets. As a result, obstetricians with private offices made fortunes by providing backstreet abortions.

When it came to morality and the sanctity of marriage, Poland was a confused place. Influenced by the Roman Catholic Church teachings, the youth – especially girls – were admonished against premarital sex by their parents, by their teachers and by the priests and nuns. Any girl that appeared just a bit off – going out with boys or dressing in fashionable clothes – ran the risk of being labelled a slut. Yet premarital sex and extramarital affairs abounded. The stigma and Church teaching notwithstanding, the general attitude toward sex was ribald and relaxed, stemming from the peasant roots of the majority of Polish society. Sex was natural, it wasn't a source of shame. And if you got pregnant, you got married. If your sex life continued after marriage, you gave birth regularly. If it didn't, you shouldn't be outraged if your husband went looking elsewhere.

A woman's experience of pleasure during the sex act was the broad undercurrent flowing through *The Art of Loving*. Wisłocka insisted on not rushing to intercourse as the one and only activity of sex life.

She exhorted love and respect, understanding and consideration. She preached that men were responsible for making sure women were ready for sex and that only then would it be pleasurable for them – a revolutionary idea. Wisłocka wanted to do away with sex that was rushed and ended immediately after the man ejaculated. In a folksy manner, she extolled the virtues and benefits of physical love. It was that vernacular that outraged the state-sanctioned sex experts: to them, Wisłocka was a dilettante, ignorant of anatomy and basic physiology, a potty-mouthed writer of unscientific purple prose.

The way I read it, *The Art of Loving* gave me permission, no, the right, to be happy with the pleasures of my body, to enjoy it and its sensations without guilt and remorse. Not to deny myself. Like any teenager, I felt indestructible. I was smarter, more progressive and sophisticated, more daring than my parents.

Two years later, already in Canada, I found *The Joy of Sex* stashed between the volumes on the bookshelf in the apartment where I babysat three-year-old Sarah. The best time to read a sex manual was at night – the yellow cone of light from the floor lamp cut a swath through the crepuscular dark and allowed my imagination to take flight to nighttime activities.

What a contrast to the Polish book. Even the paper on which it was printed – heavy, thick, almost creamy – was far superior to the grey, flimsy pages of *The Art of Loving*. The colour drawings of men and women in various postures and positions, faces glowing and flushed, tousled hair, limbs lovingly twisted around each other. In *The Art of Loving*, the chapter on positions for intercourse was illustrated with black-and-white diagrams. The man was a black homunculus, while the woman was a simple black outline: anything more detailed the government censors considered pornographic. In

The Joy of Sex, the women's excitement and satisfaction were shown in their broad smiles, parted mouths and flushed cheeks; all in full colour. The authors did not have to exhort love – it was there for all to see – and, unlike Wisłocka, they did not need to argue for a woman's right to satisfaction and enjoyment of sex.

Between these two manuals, I knew what I wanted: sex. And love. Loving sex and sexy love. I was in Canada and Polish superstitions and taboos didn't apply to me anymore. I was going to grab whatever I could and enjoy myself. But I still thought about sex in the context of love and marriage. I still dreamed about the one true love, about gifting my virginity to the man I would marry. Wisłocka's message of love and respect wasn't lost on me.

And then I met him. Twelve years my senior, a man in my mother's ESL class, he oozed sexuality like a defective microwave leaked radiation. He strutted in his cheap scuffed shoes and pants too tight around his hips and hemmed two inches too long. I had just begun to shed the twenty-five pounds I had gained after we left Austria. He looked me up and down and raised his eyebrows appreciatively whenever he caught my eye. I couldn't ignore him no matter how hard I tried. I wanted to go out with him. In Canada I had nobody. I made no friends at school and sat alone during lunches and spares. I had no boyfriend or even boyfriend material. The first time he asked me out to the movies, my father accompanied us. We saw Warren Beatty's *Reds*, a two-reel, three-and-a-half-hour bore-fest.

"Your father won't come with us again," he said the next day, proud like a hog in a mudhole. "I chose that movie for a reason." He was right – afterwards, my parents allowed me to go out with him alone. But the movie programming changed: he took me to see *Caligula* and gleefully claimed that the Canadian goody-two-shoes censors had neutered the good scenes. He had seen the director's cut in Vienna the year before. "Europeans are much more sexually adventurous," he said, his upper lip curling, and I felt daring and sophisticated. After *The Quest for Fire*, he dissected every sex scene, instructing me on what the director got wrong. *A Clockwork Orange*, the last movie I saw with him, terrified me, but I was not going to admit it. I was a big girl; I could handle it.

I was clueless.

It started with kissing in the bucket seats of his black Renault Le Car. That was all I was going to allow – no way was I going to sleep with him, especially after he told me that his goal in life was to tell his grandson that Grandpa had had a woman for every day of the year. "I'm two-thirds there," he gloated. I shrugged: I was not going to be one of this third-class Lothario's trophies.

It was exciting, titillating.

Whenever he dropped me off after our "dates" he admonished me to be careful when I got home because we both smelled like "a couple of billy goats." I would walk into the apartment flushed and wet between my legs and smirk at my parents' ignorance. I still thought I knew what I was doing. When he pleaded with me to have sex – "We'll both feel sooooo much better afterwards" – I always refused. *He* was not going to be my first. I thought I could take care of myself until a grey January afternoon after school in the vacant lot under the McDonald's billboard at Leslie Street and the 401 when he suddenly pulled his fist from my underpants and, with a sharp intake of breath, showed me his middle and index fingers. Blood around his dirt-crescented fingernails. "I'm sorry," he whispered. "I didn't mean to." As if. I remember how hard he pushed and prodded, how fast. It was exactly what he wanted.

I slept with him by the next weekend – my dreams shattered, there was no point holding out anymore. I don't remember the act, only that it happened on a creaky bed in the dank basement at his aunt's farmhouse on Canal Road in Holland Landing. He never used a condom – pulling out was good enough. He never mentioned pregnancy; instead, he insisted that if I let him come inside me, my pimples would clear up. He was so poorly educated that he thought twins happened when sperm swam up both Fallopian tubes – I had goggled my eyes at that idiocy but didn't dare contradict him.

A year later, he brutally raped me in a way I had read about in a book about Puerto Rican prostitutes, one of whom needed bowel surgery as a result. Other than the pain, and the bleach-smelling threadbare bedsheet under my cheek, and the time of the day (early afternoon) I remember nothing about it. I don't remember how I got back home; he must have driven me from the Kingston Road motel. I can't even remember if it was late spring or early fall, only that the light slanted in that peculiar way it does on this side of the equinox. After that, he vanished – no more phone calls, no more visits to our apartment. I was relieved because I didn't have to break up with him. Even then, I didn't have it in me to stand up for myself.

From then on, I knew I was damaged goods. What man with a functional moral compass would want to have anything to do with me? Add to that my Catholic upbringing, although the Polish version was not as restrictive, pure and moral as the North American one, and you have a potent brew for self-loathing. These were the subcutaneous yet fully conscious strata of guilt and moral failure as I proceeded to enjoy sex. Because enjoy it I did. Wisłocka's teachings about the joy and beauty of physical love survived my abuses; lodged into the neurons of my limbic system, not to be displaced or undone by an artless groomer, the rape and the self-loathing that followed. They endured and continued to enchant and beguile and entice me. Sex was fun and because it was fun, it was good.

Like those of many victimized women, my next three relationships progressed to the physical very quickly. My partners enjoyed my recklessness. I broke the heart of my first boyfriend, like me a recent immigrant from the Eastern bloc (Hungary), whom I met in second year of undergrad and who to this day has a soft spot for me. His replacement, a Canadian and the reason for the breakup, broke mine. With him, I took Wisłocka's instructions to heart. On our first date, back in 1984, we went to see the just-released *Amadeus*. As the shadows played on the screen, I fingered the inside of his wrist and caressed his palm. "I never knew you could do that with a wrist," he said after we emerged from the stuffy darkness into the crisp frost of a February evening. Did he sound breathless? Between the instructions gleaned from *The Art of Loving* and *The Joy of Sex*, I was ready to show him more of what I had learned. He broke up with me six months later because I was too "needy."

With boyfriend number three, I stripped naked in the front seat of his car on our first date – he was shocked but sweet enough to tell me that he liked my freckles. In spite of my excesses he, too, fell in love with me – we have been married now for thirty-three years.

Was Wisłocka ever happy in love? Satisfied with her sex life? According to Krystyna Bielewicz, Wisłocka's daughter, Staszek claimed that his needs were paramount and that she should subject herself to him whenever he wanted because his body needed sex more than a woman's. When she didn't, he began the first of his many affairs. In the hopes of curtailing his wandering eye, she arranged for her high school friend to become his lover. Wanda joined them in their one-bedroom apartment, and kept the house clean and the husband happy – a proper ménage à trois. In Wisłocka's master plan, Staszek's relationship with Wanda would be purely physical while she and he would enjoy the more superior bond of intellects. When both women became pregnant at about the same time, they all agreed that the children would be raised as twins to maintain the facade of a happy, moral marriage. Wanda and Wisłocka relocated to a remote village in Eastern Poland until both delivered, about two months apart, in the spring of 1947. Upon their return to Warsaw, the boy and girl were registered as Staszek and Wisłocka's children and named Krysia and Krzysio.[1]

Wisłocka continued her graduate studies in a research laboratory in Bydgoszcz, a city eighty kilometres north of Warsaw. She stayed away weeks at a time while Wanda remained in their apartment, officially the babies' nanny. This idyll lasted until Staszek demanded a divorce – he wanted to marry Wanda. The children were seven years old. "She was meant for sex, you were supposed to have a higher love for me," Wisłocka wailed. Then she kicked Staszek out from their Warsaw apartment. She quit her research studies and returned home. After Wanda and Krzysio moved out too, she lived alone with her daughter for the rest of her life.

In 1957, she co-founded the Planned Motherhood Society, where she specialized in the treatment of infertility and in birth control. As various contraceptives appeared in Poland, Wisłocka tested them herself before she would recommend them to her patients. She got pregnant three times and had abortions. After she finally defended her Ph.D. in 1969, she became the head of the Planned Motherhood Clinic at the Institute of Mother and Child in Warsaw and, later, of the Cytodiagnostic Laboratory of the Family Planning Society.

*

---

1 Violetta Ozminkowski, "Córka o Michalinie Wisłockiej: Kochać to ona umiała, ale nie umiała żyć," *Weedkend.Gazeta.PL*, July 13, 2019, https://weekend.gazeta.pl/weekend/7,177333,16050215,corka-o-michalinie-wislockiej-kochac-to-ona-umiala-ale-nie.html.

How could a woman whose love and marriage failed so spectacularly write *The Art of Loving*? A master of self-creation, when asked at a book launch whether the book was in any way based on personal experiences, Wisłocka famously demurred: "A blind man cannot teach about colours."

According to her daughter, the woman responsible for introducing millions of Polish women to the pleasures of physical love never found lasting love herself. After her divorce, Wisłocka had many lovers, not necessarily monogamously. She did discover sexual satisfaction, and possibly love, in the arms of a married man later in life. Jurek was a merchant marine sailor well versed in the arts of love because, as he liked to brag, he had sampled love on all five continents. But he wouldn't leave his wife, ostensibly for the sake of their daughter. In the end, Wisłocka broke off this relationship as well. When he died of a heart attack several years later, Wisłocka didn't know about it for two years.

After discovering sexual pleasures in Jurek's arms, she became supremely confident in her own sexuality: at the age of eighty-four, recovering from a heart attack in a hospital ward, she claimed that the young doctor in charge of her care was besotted with her. "There is a young female resident here who's in love with him," she told a visitor, "but she knows I have bested her, poor thing." She died two days later.

A biopic about Wisłocka could only, if unimaginatively, be titled *The Art of Loving*. I watched it on Netflix in 2017 and it stirred emotions I had not examined in years. In the final scene, Wisłocka is shown rolling a sheet of paper into a typewriter. Tears streaming down her face, unconsolled about Jurek's death, she types the title of her magnum opus and we, having watched the film, know how successful and powerful it will be. Wisłocka was gratified and thrilled when her book came out: she had won over the censorship, she empowered women to take control of their sex lives and reproduction. But was she really happy? Living her life as a single woman, with no real love and devotion in it?

What about me? Would love – and life – have turned out differently for me had I not fallen for the monster? How does it feel to have your first love reciprocated, even if only as a teenage infatuation? I found myself envying Wisłocka's life a little, no matter how chaotic and lonely it turned out to be. And I sometimes catch myself wondering how would love in my mother tongue feel. Love in Polish tempts, suggests unknown and unexperienced wonders and sensations, more visceral, more immediate, not the cerebral and remote English endearments. Would sex with Polish whispers wafting over it be more loving? Would sex cries in Polish be more guttural? Happily married to a Canadian for thirty-three years, I'll never know.

But one thing I know for sure – I have never been a slut. Or rather, if exploring and owning one's sexuality means that a woman is a slut, then I am one and proud of it.

TATO

You grew up the youngest son of a small-holding farmer – out of nine children, two sisters on either side of you didn't make it past infacy, an older sister died of appendicitis at thirteen. Your nine-year-old toothy smile in an old photo softened by the passing years is the epitome of happiness. Your clothes many times mended, your too large hunter's hat rakishly askew.

From whittled willow sticks and slices of potato, you built turbines over the stream that cut across the corner of the schoolyard. Later, you designed steam-powered turbines for coal-burning electric power plants, the backbone of the economy and the pride of Communist Poland.

In grade three, the teacher slapped you with an F in math because you refused to come to the blackboard. Like always, you had solved the math question in your head before anybody in the class had a chance, but that day you had torn the seat of your pants climbing the fence to avoid being late after recess and were too embarrassed to get up.

After you finished the two-year die-and-cast trade school, you enrolled in night school. During the day, you worked as the village mailman. As you waited for the addressees to open their gates, you solved quadratic equations from the algebra textbook you carried in your mailbag.

You missed your train stop on the way back from night school once, engrossed in *Pan Tadeusz*, the nineteenth-century twelve-book epic poem in hexameter you adored. You walked back home along the train tracks in the midnight darkness, the smell of urine wafting from the railbed. You cried when I presented you with a copy of its first edition for your sixtieth birthday.

When you were seventeen, you went to a symphony concert with the music score for the concerto being played that night. You thought you were faithfully following the score only to find yourself with two pages to go when the first movement ended. You took me to hear "The Four Seasons" when I was five and to see Bizet's *The Pearl Fishers* when I was nine. Thirty-five years later and on another continent, we still had an opera subscription.

You got into the Silesian Polytechnic on a full scholarship after you aced the entrance exams. Once there, instead of studying, every night you played bridge and chess but had no troubles with your midterms. You and your roommate invented a guerrilla vocabulary game: you would ambush each other with a word to define according to the *Dictionary of Polish Language*. Vodka and beer were ubiquitous but you had sworn not to drink because you had seen what alcohol had done to your father and your older brother.

At a party in your residence, you asked a girl to dance. Halfway through the second song, you dashed away, leaving her standing in the middle of the floor. You had set a clutch of eggs on the stove and by the time you returned, the water had boiled away and the exploded yolks spattered the kitchenette walls and ceiling.

You married the love of your life – the girl on the dance floor, my mother – in your final year of university. The vocabulary guerrilla war partner was your best man. I was born twelve months later.

After you defended your master's thesis in mechanical engineering and graduated, you took a job as a shift supervisor at the Blachownia power plant. Two years later, you were allotted a two-room flat in a town nearby. The apartment of my childhood was in a five-story tenement building that stood next to a slag-covered playground with rickety swings and a half-filled sandbox. From our kitchen window, I saw the smokestacks of the power plant – your power plant – belch with steam.

When I was burning up with measles, a pinpoint red rash all over my body, you bought me a stuffed toy beagle. We called him Huckleberry, after the character from Hanna-Barbera cartoons, the only ones shown on Polish TV that broadcast in black and white.

On her birthday, my mother lost her temper with you because you presented her with a bag of onions. She wasn't patient enough to find the long amber necklace secreted inside, its sun-filled stones perfectly camouflaged by the onion skins. By that time, you were drinking a beer or two every night.

Once you read *Winnie-the-Pooh* to me, the chapter where Pooh and Piglet build an elephant trap. We both laughed hysterically at

Piglet squealing "Helephant, helpalanth" as he streaked away from the barrel-headed Pooh and the memory still brings a smile. Greek myths were the other reading staple.

You bought *A Stroll in the Depths* for me even though I couldn't read yet. I had begged and whined in front of the bookstore because I loved the cover: a clown fish hovering above a hermit crab in a murex shell under a dome of turquoise-blue waters, the colours such a contrast to the greys and beiges of our everyday life. To this day, I love marine biology and I still have the book.

On Sunday afternoons in grade one you taught me set theory using animals of the savannah and the jungle as set elements. Venn diagrams, Boolean logic notations a new alphabet, your blue eyes twinkling at my comprehension and eagerness to learn.

One day, you wrote a list: Luxor, Karnak, Thebes and explained that they were archaeological digs in Egypt you would take me to after I passed the *matura*, the final exam at the end of high school. "We'll go to Greece, too, see the Acropolis and Delphi," you said. My mother sneered when I proudly showed her the sheet of paper.

When my new grade four home-form teacher recommended I skip a year, you beamed with joy and pride. In my old school, the teacher had refused to even consider it.

"I'm going to Canada in search of cake, not bread," you announced as we packed for our illegal emigration from Poland in 1980. In the famous nineteenth-century Polish novella *For Bread*, a destitute Polish immigrant jumps off the pier in the New York City harbour after he finds out his beloved daughter had prostituted herself to pay their rent. My mother had only agreed to leave Poland because you promised to quit drinking.

The recession hit Canada in 1982 and you, too proud to beg for a job, decided to drive a teal-and-orange Beck taxi full time. You drove nights, because the fares were higher, and weekends, because there were more fares to be had. But every Sunday you woke up in time to drive me to Robarts Library for its 1:00 p.m. opening so that I could study there all day and all evening.

You gave me one American dollar bill when I was accepted to medical school. "This is all I can give you right now," you said, "but I am so proud of you." For graduation, chuckling, you handed me a ten-dollar bill. American, of course. I still have both of them, tucked away in a fat pediatrics textbook.

When a massive heart attack felled you just before your fiftieth birthday, I thought you would never see me succeed. But that was only the first time you cheated death.

When I graduated from medical school eight months later, you dashed around King's College Circle with your camera, your aneurysmal heart notwithstanding. You didn't miss a single building, shot the entire thirty-six-exposure film: "So that you will remember what the university looked like the afternoon you graduated."

You never pressed me for grandchildren, never even asked why I didn't get pregnant for seven years after I married. When bothered by nosy friends, you would answer: "Far be it from me to tell a pair of doctors how to make babies." You spoiled both your grandsons when they arrived but never failed to admonish them to "be patient."

Ten years later, after your first esophageal hemorrhage – from the alcoholic liver disease you had developed after years of drinking ten beers every day – you told the intern looking after you that if she listened very, very carefully, she'd hear "I love you, I love you" when she put the stethoscope on your heart, you old flirt. You had no idea that you had cheated death again.

Every December 31, my birthday, you brought a bouquet of flowers. You drove fifty kilometres to my house first thing in the morning to be there for 10:15 a.m., the recorded time of my birth.

In 2009, five days before you fell and hit your head, we attended the Canadian Opera Company's premiere production of *Simon Boccanegra*, a Verdi opera about a betrayed man who, on his deathbed, forgives his enemies and reconciles with his only granddaughter. It was one of your favourites.

Even though I know you were unconscious, hemorrhaging into your brain, I can just hear you admonishing the ambulance driver: "Don't turn into that churchyard, I'm not ready yet," as he raced with your body, lights flashing, to the Milton District Hospital emergency that April evening.

You would have turned seventy half a year later.

I miss you.

A MOTHER-DAUGHTER PHRASEBOOK: DEFINITIONS

Love /ləv/[1]

"Your hair is so thick," you say. "Make sure you brush it every day." Lying across your lap, your fingers in my hair, I am surprised by the lightness of your touch – such an unusual feeling. The smell of your body wafts from underneath your skirt and I feel the warmth of your firm thighs against my three-year-old chest. I ask you to finger my scalp over and over again as time stretches toward the evening.

On December afternoons when dusk falls at four, you peel and grate carrots, spoon white cane sugar into them – I love the extra sweetness. After I gobble up the mush, I demand applesauce and you shred a peeled apple on the same grater, its plastic forever stained pale orange. Sugar on the applesauce, too. To buy crumbly milk chocolate and unripe Cuban bananas for me, you stand in lines for hours – an everyday occurrence in Communist Poland.

---

1   As per *Oxford English Dictionary*, the pronunciations given are those in use among educated urban speakers of standard English in the United States.

Fear /fɪə/

I run up the stairs to our fourth-floor apartment, my white socks caked with the oily forest loam, my breath rushing in and out of my chest. I lug an armful of pussy willows I have snapped off the branches that trail in the stream cutting through the dense shrubs behind our building.

"Don't play in the forest," you say when I burst through the door to our apartment. "There could be rapists in there."

A menace lurks in your voice and, from now on, it will forever await me behind the tree trunks sticky with resin, in the patch of wild raspberry canes in the forest, in the tall grasses whispering in the wind. What are rapists? In Polish, the word rhymes with "tormentors," a word I do understand. My pulse thumps beneath my belly button, but not with exertion anymore. I am five and I am terrified.

Care /kɛ(ə)r/

During the war, alone after your father was imprisoned in an POW camp, your mother sold cigarettes smuggled across the San River from the German-occupied territories, an offence punishable by prison or concentration camp. When that wasn't enough to ensure you didn't go hungry, she caught pigeons and cooked broth for you. Every nanny I have is subjected to obsessive questioning when you return from work in the evenings: "What did Małgosia eat today?"

When you were four, right after the war, your mother knit a blue sweater for you, bluer than the blue sky. For me, you buy hand-knit dresses made to order and a rabbit-fur winter coat – I have to look better than everybody. My grade one school uniform is tailor-made. For school, I have all the books, pens, crayons and paints, notebooks and a leather school bag. I always have good shoes – did you? Were you cold in the winters?

Your mother died when you were eight.

When I'm a teenager, you knit sweaters for me with thick, richly coloured wool: one – oversize, knit in splotches of beige, white, brown and just enough pink to make all the colours sing – I still cherish.

Family /ˈfæm(ə)li/

Your sister Danusia is less than a year old and you have just started school when your mother dies of tuberculosis. Antibiotics reach Poland too late to save her. For fear of infection, your mother is never allowed to hold her newborn daughter. Danusia is adopted by your married uncle because your father couldn't possibly look after such a small baby. Throughout your childhood, you watch them dote on her – your father and his brother shared a communal apartment. Danusia has pretty clothes that arrive in parcels from America, a belly full of food and all the love and kindness and attention a child needs. Not even a year after your mother is buried, your father marries a much younger woman and she and her baby boy keep him busy even though she turns out to be fickle and a flirt. Three years later, your father decides that the brown-eyed runt of a boy looks nothing like him and divorces his mother.

How often do you go hungry after your mother dies?

Hunger /ˈhəŋɡər/

One afternoon in the well of the inner courtyard of your build-ing, five-year-old Danusia offers you her afternoon snack – a slice of bread thinly spread with butter. You taste the butter from the UNRRA package – so sweet, so smooth on your tongue, your lips shiny with it, when Danusia's adoptive mother yanks your braid.

"You thief." She grabs the bread out of your hand and shakes you by the shoulder. "I didn't make it for you!" she shouts.

Did your father beat you this time, too?

Did he do anything else to you after his divorce?

Ambition /æmˈbɪʃən/

When I am in grade one, you supervise my homework six days a week and insist I finish all the arithmetic problems in the textbook even if the teacher doesn't assign them. When I do, you write out new ones for me to practise. Same with letters and spelling. You drill and drill me for hours at a time, tell me about your mother's beautiful penmanship, so much better than my chicken scrawls. I have to be the best at school. And at piano, art classes, figure skating, ballet. You walk me to all the classes twice a week, and for skating practise, we ride the bus to the next town over. After the homework is done to your satisfaction, you sit with me at the white-and-black keys of the piano, the metronome ticking over my head. Scales, études, bagatelles – I practise and practise. You never had the chance to learn to play so I have to.

You never encourage or praise me, but you expect excellence in everything I do.

Reprise /rəˈpriz/

My sister is born when I am nine, only a year older than you when you lost your mother. I have begged for a sibling for years. I love feeding her, marvel at the strength of her suck when I slip my pinky into her mouth. You catch me once and yell that my finger is covered with deadly bacteria. I iron tiny muslin shirts and cotton diapers, wheel my sister in the pram whose maroon hood stands almost as tall as I do.

In the mornings, I leave for school without breakfast. You

doze on and off in the rumpled bed, the crib at its feet, not getting dressed until I get back from school, your hair uncombed, dark roots widening with each passing week.

I think that with the birth of my sister you re-enacted your own abandonment; your postpartum depression a mourning for your own mother.

Malice /ˈmæləs/

"Don't you realize that the only reason Mrs. A. wants you to skip a grade is because she's tired of your shenanigans?" you ask when my father hands me an upper-grade geography textbook, his eyes twinkling with pride. School textbooks were scarce in Communist Poland and he has managed to get it from under the counter.

I'm nine, yet the illogic in your statement stuns me. I furrow my forehead, try to understand: If Mrs. A. wanted to punish me, she'd move me to the parallel class, wouldn't she? But I say nothing. Is this my first inkling that something is amiss?

You never use any of the diminutive forms of my name and there are at least fourteen in Polish.

I don't remember you ever kissing me.

Ridicule /ˈrɪdɪˌkjul/

"Do you have any idea how that has just stretched your stomach?" you say after I devour a butter-and-tomato sandwich on a kaiser bun. I am ten. I shrug – it tasted so good, the salted tomatoes still tangy on my tongue, who cares what my stomach is doing. I roll my eyes behind your back.

"You look like a cow; you shouldn't ride a sports bike," you say after I jump off after a race around the block on a friend's bike. I imagine my huge butt hanging off the sides of the bicycle seat and

I burn with shame. I am twelve and this time I care.

When we arrive in Canada, I am overweight from the dozens of Lindt chocolates I devoured during our six-month wait for immigration papers in Austria: a cornucopia of sweets I had never seen or tasted before in my life. You smirk every time I lumber toward the fridge; my skirt, held with a safety pin where the button had popped off, digs into my waist. I have a double, possibly a triple chin, you tell me. When I scowl at you, you declare that I have no sense of humour. I am sixteen.

Confusion /kənˈfjuːʒən/

You thrust the just-published *The Art of Loving* at me when I am thirteen: "Might as well read it," you say. No, not Ovid, but the first sexual education manual to appear in Poland, its pale pink cover sporting a drawing of the back of a demure bridal couple, she in a white veil topped with a virginal wreath. Communists mustn't appear to be promoting the Western debauchery of premarital sex.

Inside, line diagrams of flaccid, semi-erect and erect penises; a map of male and female erogenous zones; several line drawings of intercourse positions all interspersed with homey advice on foreplay and contraception, and disarming exhortations on love and respect.

A week later, "You slut!" rings in the evening-darkened stairwell of our apartment when I return home. A slap rings out as it lands on my left cheek. "Always a quick little learner, aren't you?" you hiss. I have no idea how you know that I have been kissing a boy for the first time behind our apartment building.

My cheek stings as I slink into my room.

Envy /ˈɛnvi/

All through my childhood, I hear you talk with admiration about other women's fur coats, patent dress shoes, gold rings and coral necklaces, leather gloves; berate my father about others' lakeside cottages or gardened city homes with wallpapered walls and Persian carpets. Foreign cars and telephones with dial buttons. Other husband's higher salaries and business trips abroad. Apartments that have more rooms and are on a higher floor than ours. No matter what my father provides, it is never enough.

Immigrating to Canada is meant to satisfy you, but recession hits the year after we arrive and my father goes unemployed for four years. You never let him forget it. You threaten to pack up my six-year-old sister and go back to Poland every chance you get.

Shame /ʃeɪm/

"Don't you know I could seduce him away from you if I only wanted to?" you ask, smiling sweetly over steaming barley broth, my favourite soup, just the two of us at the dinner table. For weeks now, you have been preening in front of a third-rate Lothario you have invited for dinner from your English as a Second Language class. In Poland, a character like that would have never been allowed in our house, and I never thought that he could be a source of envy for you, a married woman and a mother. I am seventeen, he is thirteen years my senior, yet you have been allowing him to take me out for a while now. In a delusion of teenage invincibility, I am sure that I can handle myself; like the nice girl that you want me to be, I know my husband will be my first.

I am rendered speechless by – what is it? Delusion? Pathological envy? Thirty years later, I still lack words to categorize this viciously inappropriate statement.

In a few months, your Lothario will rape me in the Pine Court Motel on Kingston Road and I will never tell you or anybody about it.

Resentment /ri'zɛntm(ə)nt/

"Keep your precious English to yourself, you selfish brat," you say and storm out of my room after I struggle a moment too long to find the English equivalent of *adwokat* – "barrister." I was trying to come up with the word, the wheels in my mind whirring and coming up empty. My eyes widen at your unfairness.

I am eighteen and have just started university. There has never been any doubt that I would continue my education – even when my father was unemployed and then driving a cab. But you . . . after you failed your final exams in high school, your father said you were educated enough. You trained as a draftswoman but when you met my father you lied that you were studying medicine. Twenty-two years later, it will be I who becomes a medical student at the University of Toronto.

Inheritance /ɪn'hɛrɪtəns/

"An intelligent person is never bored," a piece of wisdom you bestow upon me time and time again. You are never bored. You work overtime both in Poland, lugging home cardboard tubes with architectural drawings you render in ink on our kitchen table after I go to sleep, your blond hair shining in the cone of a tabletop lamp; and in Canada, where you hustle furniture at Sears every weekend no matter how many times I ask you to spend time with my sons. At home, you are always scrubbing the kitchen tiles and washing your crystal stemware and china in sudsy, scalding water. You run the laundry twice weekly and iron all your clothes – appearances, always appearances.

I inherit your work ethics and, in spite of your repeated proclamations, I am not lazy. But I do allow myself to be bored and don't feel any less intelligent for that.

Womb /wum/

You need a hysterectomy. Fibroids the size of oranges have proliferated in your uterus. The bleeding leaves you tired; your belly's bloated, aching. *So where I came from will end up in a jar on a shelf in an anatomy lab, a grey blob floating in murky formalin* flashes through my head when you tell me about your scheduled surgery. In sympathy, my own lower belly tugs somewhere beneath my navel.

For forty weeks, a twisted, pulsing cord bound us together, attached me to you inside your womb. You were proud that you didn't show for seven months and you told nobody you were pregnant.

The star cicatrix of my belly button remains the only sign of connection between us.

Birth /bərθ/

"They handed you to me after you were born and you were so red and wrinkly," you tell me when I'm six. "Like a flayed frog. Screeched like one, too." And you laugh – such a great joke. Years later, when I learn enough pediatrics, I am convinced that I must have been a preemie but you dismiss my theory. "I would know when I got pregnant, wouldn't I?" your voice offended as if I suggested something unsavoury.

When I am pregnant for the first time, I don't tell you until the seventeenth week. I have no desire to share the joy with you. Why would I? We never connected as women. I worry that you may spoil my joy with your snarks, your old wives' tales, your stupid comments. I finally phone you with the news only because my husband insists it's time.

At the hospital where I work, when I see women delivering babies, their mothers at their side, I always think it weird. I couldn't imagine you holding my hand, wiping my forehead with a wet cloth. And yet. When my epidural doesn't work, when unrelenting waves of pain surge through me from bow to stern, I do call out for you over and over again: *Mama, Mama.* And I feel mortified that I do – you have never been there for me, so why should it change now? How foolish.

When the iron fists of letdown squeeze my suddenly D-cup breasts, when I pump milk and worry that I don't have enough, you tell me that you never had any milk for me and that I turned out all right. For seven months, my son grows big and fat on my milk alone.

Love /ləv/

"She just *adores* her daughter," you say when a woman friend leaves after tea, the disapproval unmistakable. A mother in love with her adult daughter, what a concept. In medical school, a wise Polish-born psychiatrist explains to me that in Polish families, girls receive approval from grandmothers, aunts, godmothers, not from their mothers. You didn't have any such women in your life and neither do I. You never said you loved me. I know that you lost your mother, that you were neglected and might have been abused, and yet I still don't understand the things you did.

During both of my pregnancies, when I learn that I am carrying a boy, I am ecstatic.

I never wanted a daughter. I wouldn't know how to raise her without hurting her.

HIS AUNT MUST HAVE GIVEN BIRTH TO HIM!

"Go read to Dziadek," my aunt Adela says and hands me the local weekly paper.

It's summer 1975 and I am spending a few holiday weeks at my aunt's farm, investigating the barn and the very pregnant cow hiding there, climbing the linden tree in the yard to pick its fragrant flowers for tea and exploring the surrounding pine forests redolent with resin, creeks dappled with sun and meadows blooming red poppies and blue cornflowers. Reading to my grandfather is last on my mind.

I reluctantly take the flimsy newsprint. Its ink stains my fingers and runs onto my sweaty palms.

Dziadek Wojciech gives me the heebie-jeebies. I am eleven and he lives halfway across Poland. I have only seen him at family gatherings where he has always been crotchety and cantankerous. What's even more frightening is that after several failed cataract surgeries, he is now legally blind. How could he live without seeing after he's lived all his life as a seeing person? After Babcia died two years ago, their house and farm were sold and my grandfather moved in with one of his daughters. His wife had been his caretaker ever since his vision failed and his heartbreaking wails as he clutched her coffin in the funeral chapel still ring in my ears.

"Who's reading to me today?" he hollers from the porch where the weight of his short, stocky body has worn a hollow in the ancient armchair. He sits there all day long, rising only to use the outhouse and for meals. His fingertips are yellowed with nicotine from the thick cigarettes he rolls himself and his voice is raspy. Most of the time, his head hangs down, cocked to one side as if he were listening; he juts his chin up only when he utters his pronouncements. He has an opinion on everything.

"Go!" My aunt shoves me out of the kitchen.

I settle on a little stool by Dziadek's knees and read the headlines so he can tell me which story to read. The obligatory reading of obituaries follows current affairs. He knows a few of the people listed and regales me with their life stories, which he peppers with unabashed vulgarisms.

One day, after I read an obituary for a distant cousin, I ask him about his parents.

Dziadek blows out his lips. He mumbles about his good-for-nothing father who gambled away his rich wife's dowry and sold his son's – Dziadek's – bicycle to buy more moonshine. Almost thirty years later, I will learn my great-grandfather Jakub Nowaczyk was employed as a foreman on many estates but never stayed long in one place, dragging his growing family from village to village across Greater Poland. Family lore has it that he was very intelligent and that one manor lord told him to stop being so smart and to start working. In July 1899, he spent three nights in a Prussian jail in Poznań, the provincial capital, because of a fistfight in a bar. When he presented himself in court, he refused to answer the judge's questions claiming he couldn't speak German. It garnered him not only a charge of contempt for the court but also the fame of a Polish patriot and a write-up in the regional paper as an "oppressed Pole," a notice I found in the internet newspaper archive.

But when I ask Dziadek about his father's mother, he brushes me off. "Who knows?! His aunt must have given birth to him!" he says and I erupt in giggles having just realized where my father got his warped sense of humour.

Twenty-seven years later, I found myself obsessed by an unexpected desire to learn more about my ancestors. I had given birth to my second son and, while on maternity leave, out of sheer boredom, I began to organize my older son's photographs into an album. That soon telescoped into organizing wedding and travel photos and assorted others. When I finished, stacks of photographs and film negatives lay scattered all over the house and I was left with a cardboard box with old family photographs. I had collected them as a child, back in Communist Poland. These sepia photographs, corners softened by the passing time, of unnamed relatives rigidly posing in their best outfits, suggested a world so unlike the everyday greyness of my childhood, they had enthralled me. They gifted me with a sense of belonging.

Soon, I began to envision an album harking back to Victorian times: photos affixed with black sticky corners, labelled in perfect cursive handwriting – white ink on black pages detailing dates and places, first names, surnames, maiden names. Very bourgeoise. But I didn't know who those people in the photos were; I didn't know where and when the pictures were taken. These long-cherished photographs could not tell their stories – they lay on my dining table, mute and lifeless pieces of cardboard – as if I didn't care about those people at all.

But I did care.

In Poland, genealogy has always been the purview of the aristocracy and landed gentry, descendants of knights, princes and feudal landowners, and the historians studying them. The aristocrats did not have to search for their roots – their ancestors were already well-documented and part of recorded history. Members of the lower classes may remember the names of their grandparents, but their great-grandparents' remain largely forgotten. Peasant family trees are stunted; they may have many branches, but they break off abruptly after only two or three divisions. Their roots reach two generations at best in the shallow, sandy soil of memory. Neither cultivated nor tended, peasant family trees die off quickly.

I began mine where any self-respecting family historian is supposed to start: with my parents. They gave me the names of my grandparents and the maiden names of my grandmothers, but other names, places, dates of births, marriages and deaths lay lost in the dark abyss of the past. I believed there were no civil documents for peasants from the previous centuries and, even if there had been, they wouldn't have lasted through the two world wars.

"What a stupid idea, genealogy," I mumbled to myself one evening, stashing the photos back into their box. "Complete waste of time."

That was before I read Lisa Appignanesi's *Losing the Dead* and before I typed "genealogy" into Google.

Appignanesi is a Canadian writer of Polish-Jewish ancestry. In 1998, she travelled to Poland to search for information about her mother's younger brother who was murdered by the Nazis. In the Jewish Gmina in Łódź, she received the death certificate of her paternal grandmother. The document listed the place and time of birth of her grandmother *and* the name of her father, the author's great-grandfather. Death certificates listed the deceased's parents?

If she could discover this, so could I. The next day, I dropped letters into the mailbox addressed to three different Polish city halls requesting death certificates on my maternal great-grandmother and grandmother.

On the internet, combing through an avalanche of data (just try typing "genealogy" into Google), I discovered that Polish-Americans had been successfully researching their family origins for decades. Some of them are proud owners of documented pedigrees that go back to the eighteenth century. But the vast majority were descendants of peasants who left their native lands because of extreme poverty and hardship. I realized that meant there were documents pertaining to peasants, even the poorest and most wretched, the ones poverty forced to emigrate. How would I get my hands on them?

Answering this question was how I discovered the Family History Library of the Church of Jesus Christ of Latter-day Saints. For decades, as a part of their missionary work, members of the Mormon Church had microfilmed genealogical records from around the world, including the church and civil archives in the former Eastern bloc countries. The original microfilms are stored in the Granite Mountain Records Vault just outside Salt Lake City. The six-hundred-feet-long vault is designed and built to withstand floods, earthquakes, even a nuclear blast. It houses millions of rolls of microfilms containing billions of civil and church records. The main church library in Salt Lake City makes copies of these available to anybody around the world, not only members of the church, through the network of its Family History Libraries attached to local Mormon congregations. If I wanted, with a nominal fee to cover the shipping costs from Salt Lake City to Canada, I could order microfilms from my ancestors' parishes and read through them at my leisure in the microfilm reading room at the Family

History Library of the Hamilton branch of the Mormon Church.[1]
I wanted.

Genealogy is addictive. The first time you document a relationship with a person until then unknown to you, a person from the past, you feel an almost physical proximity. My first such epiphany (of many) happened when I deciphered the birth record of my mother's father, Dziadek Piotr, which was projected fuzzily onto the white laminate of the microfilm reader. His birth and my reading of its record were separated in time by almost a century (my grandfather was born in the year 1910, I was looking at it in the year 2001); in space by thousands of kilometres (he was born in a hamlet in present-day Ukraine, I was looking at it in Hamilton, Ontario). It was separated by three official languages (Polish, which is our native tongue; Latin, the language of the church record I was reading; and English, the language of the catalogue of the Family History Library); several political systems and numerous historical upheavals; two uprootings (his to present-day Poland as the borders changed after World War II and mine to Canada); and two religions (my grandfather's baptism was in the Roman Catholic rite, I was looking at the microfilm of his baptism record in a Mormon church). Across these chasms I sensed a new connection with that Piotr Wiśniowski I had known as a child, tall and stooped from a lifetime of work as a carpenter, with smiling bright blue eyes surrounded by wrinkles, and whom I suddenly saw as a mewling newborn. He would have been born in a smoke-filled, one-room thatched hut overrun by his older siblings and chickens, piglets if his family were better off. He would have been carried to a small

---

1   As of September 2021, all of the archival microfilms have been digitized and are available online at FamilySearch. Laurie Bradshaw, "FamilySearch Completes Massive Microfilm Digitization Project," FamilySearch Blog, September 21, 2021, www.familysearch.org/en/blog/familysearch-microfilm-collection-digitized.

white church through the stubble of the wheat fields (he was born in October) that belonged to the local nobleman and which his family tended. He would have been no bigger than my newborn son, his skin just as pink and soft, and would have been cradled as tenderly, swaddled in linen sheets and maybe a sheep's fleece against the morning autumn air. I could almost hear his squeals when the cold holy water sprinkled his forehead in the small chapel infused with incense.

No matter how distant, this mystical connection happened with every ancestor I found, even those not in the direct line from me. Unknown and unnamed, they came alive. It was as if their ghosts stood behind me as I cranked the microfilm, its frames whirring by dizzyingly, and nodded happily when I identified them, the women clapping with joy. This feeling of connection was always followed by an overwhelming desire to find out more: how they lived, how they worked, in what town they lived, in what house. Besotted, I located tiny villages on World War II military maps at the Lloyd Reeds Map Collection at McMaster University, pored over history and ethnography monographs, devoured the novels and short stories of their times. Later, I stepped into the whitewashed village churches where for centuries generations of them were baptized and walked the cemeteries where they were buried – in Ukraine, in Silesia, in Greater Poland – some well-maintained, some forgotten and overgrown by weeds and bush. To make the shadows of the microfilms real, I visited civil and church archives in the big cities they never ventured to so that I could touch, reverently, the musty and yellowed parish registers from two centuries ago. Each of them two feet long and a foot wide, those books were there when my ancestors lived – astonishing. In the registers, in black ink faded to palest of grey, were their names and the dates of when they were born, when they married and when they died. All written for me to read centuries later.

Genealogy isn't just the discovery of the bare dates of our

ancestors' lives – it unveils the mystery of the past. We feel the presence of those long gone because our ancestors live in us. The compulsive drive to find out more about them makes sense when we realize that we are really learning about ourselves: if our ancestors did not flourish, we wouldn't be here today. Knowing of them and thus knowing them provides a sense of belonging. Without them you're like a plant pulled out of the ground, left to wilt. When you feel their strength behind you, it's as if you grew with other trees: some ancient, already falling down, others young, growing taller and stronger, but all of them creating the forest of your family.

What about Dziadek's unnamed grandmother? My father was adamant that his father was born in Folwark Stare near the city of Krotoszyn in Greater Poland. The official website of the town hall of Krotoszyn yielded a map of the county that showed a district called "Stary Krotoszyn." There, I also found a Ryszard Nowaczyk who worked for the city administration and, remembering that my father used to mention a cousin Ryszard, I immediately emailed him.

That particular Ryszard was not a relative, but he turned out to be better than family, at least in the research sense. He contacted Karol Nowaczyk who also wasn't a relative – Nowaczyk is a common last name in Greater Poland – but who had been tracking our surname in the Krotoszyn county for years. Three weeks later, Karol sent me a copy of the death record of my great-grandfather Jakub from 1933. Dziadek's father. It listed the names of his parents: Jan Nowaczyk born in Przecławek and Józefa née Koperek born in Pamiątkowo – "the aunt that gave birth to him."

So much for Dziadek Wojciech's wit.

Now I knew where to look for the records of Jakub's parents, which, with luck and perseverance, should lead me to his birth record. Karol also mailed the birthdates of Jakub's six children, including Dziadek, who were all born in Folwark Stare but not the one near Krotoszyn. They were born in a hamlet of the same name in the parish of Kaźmierz, 165 kilometres northwest.

The roots of my family tree were beginning to take hold.

Pamiątkowo and Przecławek are villages nested on the opposite ends of Jezioro Pamiątkowskie, a small navy-blue lake, one of many that dot the lush plains of Greater Poland. I ordered microfilms for the parish of Cerekwica, which both villages belonged to. These would contain the registers of baptisms, marriages and deaths. When they arrived from Salt Lake City, I rushed to the reading room. There, after hours of scanning hazy shadows, eyes watering and ready to give up, I learned that my great-great-grandmother Józefa ("the aunt that gave birth," etc.) was born on February 21, 1817. She was the eldest daughter of Marcin Koperek and his second wife, Katarzyna. Searching back several decades, I found that Marcin was baptized on November 9, 1772, and that *his* parents were Józef Koper and Apolonia – my great-great-great-great-grandparents. All that remained to connect Józefa with Jakub, and hence with me, was finding Jakub's birth record: for that, I needed his parents' marriage. Then my family tree would have six generations' worth of roots.

Finding a marriage record in a parish register requires patience. A lot of patience. While the year of birth can be estimated from the age at death (plus or minus a year or two, barring any mistakes), a marriage could happen at any time (within reason) between those two dates. None of the registers were indexed. Over many evenings and Saturday mornings, I read and read and read – it was a large parish comprising several villages, years' worth of marriage records in various degree of illegibility – until I discovered that

Józefa married Józef Rybakowski (yes, a Józefa married a Józef) on November 20, 1833. They had five children and were married for almost twenty years when Józef died on September 6, 1852, during an outbreak of cholera. But that wasn't the marriage that interested me – my great-great-grandfather's name was Jan, not Józef. And none of their sons were named Jakub.

I began to worry, a worry bordering on existential dread: by now Józefa was thirty-seven years old and widowed and there was still no sign of my great-grandfather. Was I following the wrong lead? Back to reading the register of marriages and soon I was pumping the air with my fist, raising a silent cheer: Józefa remarried. On June 3, 1854, she wed Jan Nowaczyk, a bachelor from Przecławek fourteen years her junior. Jan's parents were Walenty Nowaczyk and Katarzyna, and I whooped in glee: another generation of Nowaczyks!

Ten months later, on April 4, 1855, at one in the afternoon, Józefa gave birth to a son Wojciech who died two weeks later. My heart went out to her. Three years later, Józefa gave birth to another son: Jakub. Finally, my great-grandfather. He was born at three in the morning on June 28, 1858, and, like his brother, brought to the Cerekwica church for christening. Józefa was forty-one.

I sat in front of the microfilm reader speechless with awe. In the nineteenth century and among the poorest peasants, in all my searches, I hadn't come across a woman having children so late in life. I have seen many die in labour much younger. I imagined the thatched hut with earthen floor lit dimly by the fire from the clay hearth during the night she was in labour, a sooty tin pot with boiling water hanging over the fire. The straw pallet on which she gave birth. The fecund smell of the amniotic fluid gushing from her, the metallic smell of blood. No running water, no creature comforts. For her sake, I hope that it had been a warm summer night. The physical and emotional strength she must have summoned to bring my great-grandfather into the world is beyond my ken.

She must have been one hell of a woman.

I always picture her short and stocky, with broad hips and stout waist. Physically strong from years of labour in the fields of the manor lord, her hands callused and scarred by the sickle and the sharp edges of the grasses, her arms muscular. I imagine her in the first blush of youth when she married at the age of seventeen, an innocent maiden, and older and wiser, a mother of five and a widow when she married my great-great-grandfather. I thought about her sorrow when she buried her newborn son and her joy at the birth of Jakub. I realized that I respected this never-seen, poor, illiterate, long-dead woman. I did not want her to languish forgotten. Across time, from halfway around the world, she demanded to be found and for her story to be told.

I needed to find her death record to complete it.

But no matter how many times I spun the microfilms, no matter how many hours I spent searching, I couldn't find the record of her death in the parish register. After the birth of Jakub, she disappeared like a puff of smoke blown by the wind.

It took me two years to put forward a theory: her little family moved away from the parish. If they did, could it be to Kaźmierz, the neighbouring village where Jakub's children were born twenty years later? Back to the Family History Library to fill out forms for more microfilms, back to the reading room to strain my eyes, scouring more parish registers. It paid off – Jan Nowaczyk died in Folwark Stare in the parish of Kaźmierz on June 15, 1868. The record stated that he left behind "widow Józefa" and a ten-year-old son, Jakub.

Another milestone added to Józefa's life.

But I still didn't have the date of her death. For weeks, I searched in Kaźmierz and neighbouring parishes: no death record for Józefa Nowaczyk in the microfilmed records that spanned the years 1866 to 1901. Could she have lived beyond 1901? That would have made her older than eighty-three years. Not unthinkable in our times, but in the nineteenth century, for a woman from the lowest socio-economic status, maybe not so likely. She had already proven herself to be made of sterner stuff but now she was missing in the mists of history, lost somewhere in the sandy soils of Greater Poland.

But wait. Could she have remarried? She was "only" fifty-one when her second husband died. What do you know? In the Kaźmierz marriage records, I found Józefa marrying for the third time. On February 12, 1872, at the age of fifty-five, she married Jan Dziadzin (in Polish there is a plethora of masculine names that begin with letters other than "J," but my ancestors didn't seem to know any of them), a sixty-six-year-old widower. Józefa sprung alive again.

After that, it took several hours of deciphering the death register to learn that she lived another nineteen years. Widow Józefa Dziadzin – she outlived yet another husband! – died on December 21, 1891, at 6:00 p.m. in Folwark Stare at the age of seventy-four. Her son Jakub, Dziadek's good-for-nothing father, reported her death. She lived long enough to meet her son's elder children, but she would never lay eyes on my grandfather, the youngest son: he was born five years after her death.

Józefa's story kept me spellbound for four years. From the existing records of the patriarchal society in which she lived, I cobbled together her life based on her children's baptisms; her marriages in the three stages of life: as a maiden, as a mother and as a crone; and the death records of her husbands. From the birth records of her two youngest sons, I learned when she moved from one village to another. Finding the record of her third marriage allowed me to extend her life by almost two decades.

I would have loved to have met her. She survived serfdom and cholera, seven pregnancies and three husbands, the Polish patriotic uprisings of 1830, 1848 and 1863, and even the great comet of 1881.

Not bad for an "aunt" with no name.

I have no delusions that Dziadek Wojciech would have cared about my research – he didn't seem the type. He died in 1980, three months after my father drove our family out of Poland in a brown Fiat 125. Instead, my father became as obsessed with genealogy as I was. He encouraged my sleuthing and cheered every discovery. He framed the pedigrees I drew and hung them in the hallway outside his home office. In 2003, he and I flew to Poland for the first time since we had left. For two weeks, without a word of complaint, he drove the standard-transmission Renault Mégane wherever I directed him, bumping on potholed and meandering country roads from village to village, cemetery to cemetery, church to church. He introduced me to his cousins who shared family lore. He accompanied me to church and city archives in Poznań where we found his grandmother's death record. In the civic office in Luboń, we asked to see my father's birth record and we were both amazed to see Dziadek Wojciech's signature attesting the birth of his youngest son on October 30, 1939. Over the yellow page with the faded scribble, my father and I looked at each other and smiled, and maybe, just maybe, somewhere else, Dziadek smiled, too.

WATER UNDER THE BRIDGE

"But we already said goodbye," I told Cezary on the phone.

The night before, he had dropped me off at the hotel after dinner with Asia, my best friend from high school. I marched away straight-backed, chin held up, and made a point of not looking back. My posture hid a broken heart and hurt pride – that whole evening, Cezary had treated me like an old friend. Platonic friend. In the hotel lobby, I had a childish hope that he would be following me with his eyes, sorry to see me go. I even asked the receptionist whether the gentleman was watching. No, ma'am. He had already driven away.

So much for that.

"I'd like to see you again," Cezary said, three days after our awkward first meeting.

Twenty-three years earlier, before my family had immigrated to Canada, these words would have sent my heart cartwheeling in my chest. Cezary had been my childhood crush that blossomed into a fiery first love. Unrequited. Now, hope, the thing with feathers, lifted its tiny head in my chest. Maybe if we were alone he'd admit to his true feelings, the feelings he had hidden all those years ago, the feelings rendered stronger with every day of my absence?

"My father and I leave this afternoon," I said. Of course I wanted to spend the day with him: we hadn't been alone since the first time I had seen him, since I had left Poland. He had gained weight and lost all of his hair, but his smile still dazzled, his eyes shone just as blue as when I first saw him when I was ten.

"We could go to Toszek, see the castle," he said. "It's pretty close."

There was a psychiatric hospital in Toszek; in our local vernacular "going to Toszek" meant going crazy. "Let's go see the crazies," I said.

As we met downstairs, my father, lounging on the hotel lobby couch, grumbled from behind his paper: "Make sure she comes back on time." In Toszek, we drove past the asylum – it had closed in the mid-eighties. The three-story red-brick building loomed behind a high wooden fence, staring at us with the empty sockets of its smashed windows. In the castle ruins, October-yellow poplar leaves rustled as we sauntered between the ruined battlements; the air felt fresh, the wind half-hearted. White smoke tendrils layered on the brown-black harrowed fields, an acrid scent of burning autumn leaves and potatoes lingered. Fall in Poland always smelled of field-roasted potatoes. The beige tubers, plowed from the earth, roasted in the embers and picked from the white ashes, always burned one's hands and seared one's tongue.

The thing with feathers held its breath.

But during the two hours we strolled in the park, Cezary never ventured beyond small talk: his work (an architect), his sports (windsurfing and snowboarding), his family vacations (Croatia and the Masurian Lake District for windsurfing and the Dolomites for snowboarding). When we returned to Gliwice for lunch I couldn't take it anymore.

"Didn't you know I was pretty darn crazy about you back then?" I blurted out over a platter of potato-and-cheese perogies. I had to know what he felt.

"Shhhhh," he hissed, hunching forward. He cast his eyes around. Being told by an old schoolmate about her teenage crush seemed to mortify the successful town architect. Yet I, from the moment I first saw him at age ten till the day we left for Canada five years later, had told my friends about my infatuation; my parents, my cousins, his cousins, even my five-year-old sister knew. I have never been ashamed of my emotions. Of course, at times my unrequited love would turn to anger, "hell hath no fury like a woman scorned" and all that, but even then it fronted a fervent, teenage infatuation. Its remnants littered my love life long after we emigrated.

"Well?" I asked.

He shook his head.

"But everybody knew!" Who cared who heard? He didn't even know? When I left Poland – for good, as the Communist regime dictated – he was supposed to realize what I had meant to him, regret his silence. I had always hoped that deep down he had cared. This delusion sustained me through my first abusive relationship in Canada, through fumbling my way in my limited English in the cliquey North York high school, my loneliness and isolation during those first years in Toronto. In my dreams, transported back to Gliwice, I searched for him in empty office buildings, running along long, dimly lit corridors; from high windows; I would see him turn the well-known street corner lined with tall lindens or disappear behind the newsstand kiosk we had passed every day on the way from school. I would dash out, but I never caught up with him. Those dreams continued for years, even after I was married.

But he didn't need to know all that.

"Water under the bridge," I said and smiled. My teeth hurt from the pretense, but I promised myself that Cezary would not know how much I cared about being rejected. Again.

My father and I set out at 2:30 p.m. from the hotel, much too late for Poland's northern latitudes where the sun sets in the afternoon in the fall. I drove the rented Renault, my father seated beside me. The standard transmission, the narrow and winding country roads and the sunlight that slanted at an angle more acute than in southern Ontario engaged my hands and eyes, but my thoughts tumbled like laundry in the dryer, bumping and bouncing, a big moist mess. I was tormented by Cezary's smile, embarrassed by his patronizing look, angered by the pity I thought I saw in his eyes. Dusk veiled the trees on the sides of the road. You do not know darkness until you've driven on a single-lane forest road in Eastern Europe at night. The yellow dividing line shone in the high beams. The car smelled of new leather and new plastic, the heat on high just the way my father liked it. I drove in silence, staring dead ahead.

"What did you guys do?" my father asked after we passed Koźle, about an hour into the drive.

"We went to Toszek, saw the castle ruins. So beautiful. How come we never went there?"

"There're many things that we didn't do when we lived here," my father said.

As the darkness grew, I pulled the car over onto the side of the road and we switched seats. The night air cooled my burning cheeks and smarting eyes.

I curled up in the passenger seat, deflated like a balloon after a birthday bash. Cezary was supposed to have cared. He was supposed to have been in love with me; he was supposed to pine away. I stared at the darkness outside the passenger window and breathed through my mouth, tears wetting the cheeks I didn't dare wipe. I fervently hoped my father wouldn't notice.

My father was a freakishly intelligent man of hard and dated opinions, possessed with a warped sense of humour: "I'm not going to instruct two doctors how to make babies," he would tell family and friends when they inquired about grandchildren; and a nasty, wide streak of sarcasm: "If I had worked as hard as you, I would have gotten the Nobel Prize a long time ago." I feared going on this trip with him, alone for two weeks. I worried that we would fight and argue, that he would sharpen his wit on me. But when, after years of research, I had located his grandmother's village not more than thirty kilometres from our old hometown, I wanted to visit it with him. He alone in my family took interest in my genealogy obsession. In the end, I had decided that I was a big girl and that I could take his jibes and jokes. And here we were, alone and afloat in the dark like divers at the bottom of the ocean, me sobbing over an old love and he oblivious to my misery.

"When I went to university, there was this brunette back home," my father told the dark windshield, his face up-lit by the dashboard lights. Surprised, I stole a glance at him. He had never mentioned this. "She was a clerk at the village post office when I was a mailman. She was very pretty and she knew it." I wanted to scoff, seeing right through this lame attempt at cheering me up, but I kept quiet. "We went to the movies and once I gave her a book of Emily Dickinson, but she said she didn't like poems. I wrote to her from university, but she never answered."

The car flew across the darkness in silence.

"Yes?" I asked.

"When I came home for Christmas that year, she gave me a poetry book," he snorted. "But I had already met your mother."

He lifted his hand from the stick shift and patted my forearm.

He didn't have to say anything else. We were both unloved teenagers again and it was okay to cry. He wasn't going to laugh at me.

The next day, Cezary phoned me at our hotel in Poznań: I had left my laptop charger in the hotel room. The manager knew him (everybody in town seemed to know Cezary) and wondered if he knew how to return it to me. Cezary paid for an express train delivery service. There was a red rosebud tucked in the package.

Over the next week, as I criss-crossed the Greater Poland province with my father, visiting graveyards and old homesteads, dining and lunching with great-uncles and second cousins and learning to respect and love my father all over again, Cezary and I talked on the phone at least once a day. On the second last day of my visit, he told me that he planned to go to Warsaw on the weekend for the annual Jazz Jamboree. I beamed with happiness. I had a room booked in a Warsaw hotel, alone, because my father stayed behind with his older sister for another two weeks. An image of myself, naked, sprawled under his six-foot-four frame on the narrow hotel bed flashed through my mind. But no, I shook myself, of course I wasn't going to yield, no way. I was happily married, I had children. But this time, for sure, he would declare his love for me. He would be the piner-after now. That was all I wanted.

He never showed up.

The morning before my flight, Asia phoned me when I stood at the check-in counter. Cezary had come by her dental office that morning to return the folding umbrella she had left at the restaurant – "Can you imagine a feebler excuse for a visit?" she laughed

– and he couldn't stop talking about me. Her inflection left room for no doubts – Cezary was smitten.

"You made quite the impression," she said.

I flew home to Canada giddy with happiness. Cezary cared! Finally! Tipsy from the cheap airline wine, I fantasized about innuendo-laden Skype talks, sexy emails, romantic dates in Vienna or Paris. For eight hours, uninterrupted fantasies rushed through my veins and pooled in my lower belly, my body humming. Even in the taxi from the airport, my body still thrummed like a taut bow.

At home, Jimm met me at the door. "I missed you," he whispered into my hair as he led me into the hallway smelling of fresh crêpes, melted butter and cinnamon. He had made his signature breakfast to welcome me home. Jack and Luke were stacking Lego blocks on the indigo carpet in a playroom flooded with sunlight, a maple tree standing in a carpet of red-orange leaves shining through its windows. They shot up the moment they saw me and ran to me, climbing my legs like two spider monkeys. Under their weight, I almost toppled onto the rainbow plastic shapes strewn under my feet.

The whole world reset itself. I belonged here. Nothing else mattered.

Enough, I thought. No more crazies.

My father would have approved.

DYED IN THE HAIR

If you pluck a grey hair, ten others will grow in its place.
– Polish folk wisdom

During the fifth week of pregnancy, as the forming fetus swims in its ocean of amniotic fluid, a sheet of epidermal cells beneath her skin thickens and dives deep into the subcutaneous tissues. The deeper layers curl up and cup the finger-like projections forming hair follicles. Thousands of these bulbs fatten beneath each square inch of the skin and soon lanugo – a crop of fine hair that falls out during the last trimester of pregnancy – covers our girl. By the time she squeezes through the birth canal, her entire body – except palms and soles – is covered by fine, much shorter fuzz. In contrast, her scalp, eyebrows and eyelashes bear hairs that are both thicker and coarser: the terminal hairs.

*

At the age of one, family photographs show me with wisps of white hair plastered onto my skull as I chew on the crib rail or cheekily wrinkle my nose at the camera. My mother's locks were chestnut brown, my father's straight tresses pitch-black, but both sets of my grandparents claimed my parents were blond as babies.

By the time I turned five, I was the proud owner of two shoulder-length, mousy brown pigtails my mother braided painstakingly every morning before securing them with an elastic cut from a punctured bicycle inner tube – no fabric-wrapped elastics in Communist Poland. She tied big satin bows over the elastics. Fire-engine red, pink, white, royal blue – we had plenty of those. On school holidays and on Sundays, my mother added two more at the top of the braids.

All my childhood, my hair was never cut. I waited for my hair to get long, long like my friend Beata's whose braids reached to her bum, but my hair never grew beyond my shoulder blades. Instead, with age, it grew relatively shorter, until my pigtails barely touched the tops of my shoulders.

As the keratinocytes – the cells at the root of the hair follicle – mature and multiply, they push older cells away from the blood vessels that supply the skin with nutrients and oxygen. Starved, the older cells die and become keratin, the protein of the hair shaft – the hair is basically a graveyard of cells. The dead keratinocytes are compressed to form the uniform core of the hair, and as they are pushed upward by the constantly dividing younger cells, the terminal hair emerges. It grows longer as each generation of cells dies. The shaft's final length depends on the tensile strength of the keratin molecules: the softer the keratin, the sooner it breaks.

*

Saturday is hair-wash day. While I soak in a deep bathtub, my mother shampoos my hair twice. Once for getting rid of the dirt and once more for shine and lustre – that's what Polish beauty books advocate. And, since conditioners don't exist in Poland, as I climb out of the tub, my hair rinsed squeaky clean, I dread what awaits me: the combing-out-of-hair-after-the-wash. It hurts. Un-moisturized and dry, my fine hair tangles like a Gordian knot and every Saturday I beg my mother to let me cut it.

"Short hair is for boys," my mother says as I squirm and hiss and ouch with every tug for what seems like hours until she can finally run a fine-toothed comb through my wet tresses.

The keratin core of hair is wrapped in hair cuticle, a very thin layer of dead scale-like cells. These overlap each other to resemble roof shingles whose free edges point upward. The scales of the cuticle are moisturized by sebum, a waxy substance secreted by the glands at the base of the hair follicle. When the sebum is washed off by the shampoo's detergents the shingles of the cuticle pull away from each other, their edges exposed. It is those edges that are smoothened by the waxes of hair conditioners and the silicone of hair detanglers. When the cuticle scales rise up or crack off, the hair frizzes and breaks easily; when the scales hug the hair core tightly, the moisture trapped inside the hair shaft renders the hair lustrous and shiny.

"Her mother must be dyeing her hair," the busybodies sprawled on the bench in front of our five-story apartment building whisper loud enough for me to hear as I pass their observation post. They sit there every afternoon, crocheting blindingly white lace curtains and tablecloths from cotton thread, spiders spinning their gossamer art. I dart into the stairwell and cry on the concrete steps until my mother finds me, snotty nosed and hiccuping, on her way from work. At the age of four, my hair is acquiring its colour – a rich, chestnut brown touched with golden highlights – in stages. Blond stripes alternate with the dark browns and for a year I am that "zebra girl" from Building Eleven.

The proportions of black and brown eumelanin and yellow pheomelanin determine where on the infinite spectrum of the human hair hues one's hair colour lands: blond, brown, black or red. Melanins are synthesized from the amino acid tyrosine by melanocytes, a clutch of weirdly spiky cells that have migrated from the nervous system to live just beneath the hair follicles. Their octopus-like tentacles creep into the crevices between the keratinocytes and deliver their pigment – packaged into tiny vesicles called melanosomes – right on the doorstep of the hair follicle. As the keratinocytes gobble up the melanosomes they acquire pigmentation themselves. When they die and transform into keratin, the colour stays within the protein. Without pigment, the hair core is colourless and appears white or grey.

After we escape Communist Poland, we live in Austria for six months. There I discover Lindt chocolates and gain thirty pounds, which I lose during our first year in Canada a year later. I begin to look good, but the same cannot be said about my hair: it breaks and splits; one evening, as I comb out my French twist, I am shocked to find a trifurcated end. I still don't use conditioner because I think that they are simply a money grab by unscrupulous capitalist cosmetics companies – my hair hasn't needed it for sixteen years, it does not need it now. My hair continues to fall out in fistfuls and one day I return home from Fairview Mall holding a sheared, limp braid by its tail.

Even my mother agrees that the pixie haircut is an improvement over my frizzy hair that resembled a heap of hay after a windstorm had its way with it.

The hair follicle spends most of its life cycle in the growth phase during which the hair shaft lengthens. Scalp hair grows for two to eight years before the follicle enters the rest phase; the eyebrows and the eyelashes stop after about five months. The rest phase lasts months to years before the hair finally falls out and the follicle involutes.

Every day, fifty to one hundred hairs are shed from among the millions of scalp hair, an alarming sight at the bottom of the stall after the morning shower, but that hecatomb of hair is nothing compared to telogen effluvium, the hair loss caused by rapid weight loss, poor nutrition, chronic illness or the hormonal upheavals of pregnancy. Up to seventy percent of scalp hair can be shed in a matter of weeks. Mine was most likely caused by the stresses of immigration, of being mute in a new language, of culture shock and of the loss of twenty-five pounds of puppy fat.

A wavy silver thread over my right brow catches the light in the mirror at the Bloor Street Fairweather store as I try on a poppy-red cotton sweater. I snake my finger through the mane of hair that falls over my forehead and hold it between my fingers.

"Is this a grey hair?" I ask the salesgirl in mock horror as I lift it out into the air. She stands speechless as if embarrassed to confirm its presence.

I am twenty, I am pretty, one grey hair changes nothing. I yank it out right there and laugh as the girl flinches.

*

Over the next eight years, as more and more of my hair follicles stop producing melanin, my boyfriend-then-husband tells me that he can't wait until I am completely grey. The first time he says it, I lie on a bench on the Seine embankment on the Île Saint-Louis in Paris, my head nestled in his lap, during our pre-wedding trip. "You have a whole crop of them in the back," he adds, fingering my scalp the way I love. I demand that he pull them out and we spend the next hour hunting down the grey amongst the strands of my chin-length bob. Thick and wavy like the coat of a Newfoundland dog, my hair is healthy and shiny, and we both think that one day it will simply turn uniformly white. We pluck out sixty-seven white hairs.

I admire women who rock their greys styled in the latest fashion. I love the long silver tresses or the pert bobs of women who remain beautiful and elegant in spite of their age, the Helen Mirrens of the world who don't give a damn. In my thirties, I begin to imagine myself with an arty white 'do à la Andy Warhol.

But over the next two decades, my hair refuses to cooperate. It goes grey slowly and untidily. No exclamation of style, only a screech of age. Soon, there is too much dull grey – no bright snow-white in sight – to tolerate and I begin dyeing my hair with a semi-permanent dye. I feel righteous; I fool myself that I am above vanity and the pedestrian need to appear youthful forever because I don't mind showing my partial grey now and then. The added benefit of semi-permanent colour is that as the dye washes out, it does so evenly along the length of the entire hair, without the telltale skunk stripe at the scalp.

Semi-permanent dyes use little or no hydrogen peroxide or ammonia to soften the hair cuticle and, as a result, penetrate the hair shaft only partially. The final colour of each strand of hair depends on its original colour and its texture, and on the porosity of the hair cuticle. Because grey or white hairs have a different starting colour than the still-pigmented surrounding hairs, after colouring with semi-permanent dye they will not appear the same shade as the rest of the hair. The final result is a more natural overall colour. I feel I am cheating nature.

"Is this a permanent dye?" I wrinkle my nose at the ammonia reek wafting from the greyish-blue goop the hairdresser spreads onto my scalp at the new hair salon I have decided to try.

"Yes," she answers in that haughty I-am-an-unappreciated-*artiste* tone affected by colourists at high-end salons.

I am not that grey! I want to shout. I still have a long way to go. And why would she assume that I wanted permanent dye?

"I didn't ask for that," I say. "I use semi-permanent dyes."

She shrugs.

I rein in my anger: she is literally holding the future of my scalp in her hands and the power imbalance is palpable. If I piss her off, I may walk out of here with no hair at all – I have heard horror stories of permanent dyes gone bad and I clam up.

And thus I enter the land of the permanently coloured.

Permanent hair dye has three components: an organic solvent, a coupling agent and an oxidant. Before the dye is applied, ammonia is used to lower the pH of the hair shaft to the acidic conditions necessary to open the cells so that the dye can penetrate deeply into the shaft. The chemicals used – toluenes, phenols, benzenes – have been shown to have carcinogenic and mutagenic properties, at least in laboratory mice and rats. During hair dyeing, the scalp becomes a laboratory bench where the ongoing chemical reactions are kept just this side of deadly. As the chemicals react with each other they produce large, pigmented molecules that enter the hair shaft. The final wash with toner – another chemical soup – seals the cuticle shut, trapping the dye molecules inside.

Any of this should suffice to scare any sane person away.

Fast-forward fifteen years. I am living the constant root-induced worry: one week after colouring, all is well; ten days, is some white showing? Two weeks, yes, the grey at the roots is peeking, but not so much; three, the roots have grown to about four millimetres and I can still pretend they're not there. Four weeks – damn! I forgot to get a return appointment and the hairdresser cannot take me till next week by which time I will be exposed for the cosmetic fraud that I am with the telltale roots shining among my dyed tresses.

Just like the melanocytes of the hair follicles decrease their melanin production, so do those scattered throughout the skin. With age, the skin loses its healthy glow, pales and, against a solid mass of invariant colour that hair dyes impart, the face begins to look ghostly, washed out. Naturally grey hair looks much better on older women. That's why so many of us dye our hair lighter and lighter as we get older, all the way to blond.

I have never wanted to be a blond. Time for highlights.

The hair strands to be highlighted have to be stripped of their colour first. Hydrogen peroxide, an oxidizing compound also used as rocket fuel, is applied to bleach the natural colour from the hair shaft providing a blank canvas for the application of the dye. After strands of hair are separated from the mane with a fine-toothed comb, the hairdresser applies the peroxide-dye mixture strand by strand and wraps each separately in a sheet of aluminum foil. Depending on the thickness of the hair, applying the foils – as per salon parlance – alone can take up to an hour. The time needed for the colour to develop, on the other hand, depends on the hue of the dye and the remaining colour of the hair: give or take forty minutes.

By the time I am almost fifty, I am tired of the whole getup: half an hour to apply the foils and the lightener; an hour of waiting for the dye to take during which I grow convinced that any second the foils will start transmitting messages from outer space; rinsing with a mordant to tint the lightened hair and waiting – again – before the final wash and conditioning. All that before the first snip of scissors! Add the travelling time to and from the salon and poof! three hours gone from my life. To look good all the time, I need to do this every four weeks, but I never manage it, so frequently I touch up the roots at home and splatter the dye all over my bath-room mirror, counter and walls – a colouring technique I highly do not recommend.

I decide to see what my natural hair colour looks like beneath all that chemical laboratory.

Science is yet to discover the exact biological mechanism of grey-ing or, as the medical literature calls it, canities, a word not even listed in the *Oxford English Dictionary*. Is it the exhaustion of the synthetic potential of the melanocytes? The accumulation of oxi-dative damage caused by free radicals released during years of hair growth? Lack of trace elements such as copper, zinc, selenium, manganese? Age-related loss of melanocyte stem cells beneath the hair follicle? Genetic and biochemical research of greying continues, funded by ridiculously huge amounts of money from cosmetics and pharmaceutical companies but – to date – there is no consensus on its exact cause.

I feel brave and daring when I make the appointment for my next haircut. Cut only, no colour, I tell the receptionist.

The young hairdresser, visibly appalled, consults the photo of a feathered, short haircut I have brought and cuts my hair in an almost, but not quite, entirely different way. I wanted to be arty and bold; instead, I look frumpy and unkempt, my hair bisected horizontally half brown, half dull grey, the look I have always dreaded. After years of chin-length bobs, I am so stunned to see myself in a matchstick-length hair that I actually tip her.

Three days later, I march back into the salon and demand a redo. An even shorter haircut uncovers the swaths of light grey on the temples that accentuates my cheekbones and broad forehead. Dyed spikes remain on top but at least the haircut is cool and the grey has finally become a statement and not simply unkempt roots.

Two weeks later, one more haircut to get rid of the coloured tips and surprise! I. Look. Good. My natural hair colour – partly grey, partly dullish black – matches my aged pale skin much better than any dye from a tube ever could.

I feel bold in that I-don't-give-a-shit-what-you-think-about-my-looks-and-age way. I still slather on foundation and eyeliner, dab on rouge and lipstick except now it is the pink shade that didn't quite do before. I feel spunky. I feel sexy. I'm fifty and I'm embracing it, cheekily. I flirt with an old friend and when he rejects my advances, I do not attribute it to the grey but to his total lack of taste in women.

I buy an industrial-size bottle of purple shampoo to keep my grey from turning yellow and brassy, and another of conditioner formulated for grey hair. I plan to be doing this for a while.

But somebody isn't happy. After offering half-hearted compliments brought forth by the initial shock, my husband begins to voice complaints. He even claims that he never – never! – wanted me to go grey. When I remind him of his sexy silver fox comments of yore his mouth drops open followed by a vehement denial. My two teenage sons follow suit – they both whine for a mom with long, brown hair, like the one they remember from their childhood. After eighteen months of this cacophony of dissatisfaction from the men in my life, I wave a white flag and I make an appointment with a colourist. Here we go again.

In a decade or so, I might try going grey again. Or I'll dye it plum streaked with carrot.

ALMOST PERFECT

"Tell me," Meghan says, looking up from Adèle's eyes into mine. "Tell me what it means." She brushes the yellow curls from Adèle's forehead and hugs her tighter, nuzzling her face in her baby's neck. As the rag doll head flops back Meghan supports it with her hand. George, Adèle's father, his arm around his wife, closes his eyes. I wonder what he is thinking.

And so it begins. Again.

The short, pert chromosome 15 unfurls itself and shakes out its folds and kinks. It and its forty-five cohorts are getting ready for their big time: ensuring survival of their genes for another thirty years or so. Duplication for the next generation. The sequence in every one of them has to be copied perfectly to carry the genetic blueprint to build a healthy, robust body. No errors allowed: the DNA repair machinery will check and recheck the copies, running back and forth along the nascent strands, scanning for mistakes.

In this case, however, about thirteen hundred years ago, amidst the ruins of the Roman Empire, one particular DNA polymerase trips, stutters and spits out a four-nucleotide sequence twice. The ligase whooshes past and stitches the duplicated fragment into a short exon of a smallish gene tucked away two-thirds down the chromosome. The repair machinery misses the insertion, as tiny as it is and easy to overlook. The sperm with the faulty gene survives, its beating tail propelling it happily up the uterus and the Fallopian tube and through the cell membrane of the egg rolling down to meet it; it beats out millions of others as it wiggles through and fertilizes the egg. The round zygote divides and divides forming a lovely thriving embryo, the error reproduced in all the cells of the healthy girl-fetus. Her own eggs begin their division well before she's born, half of them tainted with the insertion.

The affected gene codes for a crucial cellular clean-up protein, which mops up and breaks down used fats and lipids. Without it, when there are two copies of the affected gene and no healthy ones, cells choke up on their waste and die, full of abnormal metabolites. The neurons are affected the earliest: the brain, swollen and fatty, fails in its attempts at normal development.

This girl is born screaming; feisty, she wriggles in the grip of the midwife. Her cells have one copy of the altered gene, she'll be fine because the other copy suffices to do all the work. She survives without any mishaps so common in those days: she encounters no mumps virus to render her sterile, no lethal plague bacteria; she doesn't fall into a well, she is not kicked in the head by a horse. She grows up and marries a young man with soulful brown eyes, chosen for her by her parents. Half of her beautiful children – curly-haired, pink-cheeked girls and tall, broad-shouldered boys – carry the altered gene. A quarter of her grandchildren do as well. By now she and her Jewish brethren have reached the Rhineland on their northward migration from the foothills of the Alps. She has many grandchildren, her grandchildren have many grandchildren, her grandchildren's grandchildren have many more again. Some stay behind and become the Ashkenazim, "The Jews of Germany," others move East, as far as Poland, where in the fourteenth century Casimir the Great grants them special privileges and protects them as "people of the king." They prosper. They marry and multiply.

But along their travels, in almost every generation, there are wailing mothers whose infants never smile, whose toddlers never toddle and whose children die young – deaf, blind and paralyzed, choking on their own saliva or starving to death because they cannot swallow; and fathers who stop smiling as they lose son after son and daughter after daughter. The couples who bury their children are the ones where each has a copy of the mutated gene, both of them the progeny of the Italian girl – the rouge gene's mate also needs to be mutated to wreack havoc in the developing brain. The couples' other babies do well, but with time more and more of the girl's descendants carry the killer gene and more parents lose their children.

And so it goes. The gene sneaks through seven centuries, hitching a quiet ride in the recesses of the genetic code, occasionally raising its ugly mutation in a conflation of sorrow and loss. Some of the girl's descendants immigrate to the New World, but they cannot outrun the child-killer. For those who stay behind, the gene becomes the human smoke emanating from crematoria, killed instead of killing. And then it lies dormant for three more generations.

Until one spring Tuesday morning in the beginning of the twenty-first century, in my office flooded by sunshine streaming through the windows, I look across my desk at seven-month-old

Adèle with blue eyes and dark lashes curved back to the eyelids. She lies placidly in her mother's arms, a Mona Lisa smile on her raspberry lips. She doesn't wriggle, she doesn't babble, she looks straight ahead without fixing on anything – a peaches-and-cream rag doll. Adèle has been referred for a genetic consultation because she does not reach for toys, does not hold her head up and had stopped cooing about a month ago. I can tell that Meghan knows that something is wrong: Adèle is her third child. George, Adèle's tall and broad-chested father, his arm around Meghan's shoulders, takes his eyes off Meghan only to look at me like a man without a future.

When she was on the examining table, I asked Meghan to hold onto her as I reached for the ophthalmoscope to examine her eyes. "She might roll off the table," I said.

"If she rolled, I would jump up and cheer," George said.

The cherry-red spot with its milky halo at the back of her eye told me all there was to know: the past of Adèle's people and her own short future. As my heart beat wildly I finished my examination, even though there really was no reason to – all I needed to know was in the back of her eye. Still, my impression needed to be confirmed by an ophthalmologist – I could chicken out and delay telling Meghan and George what I saw. That would be callow; you can't cheat fate.

I have seen it so many times and it never ceases to shock me. I know what I have to say and I want to delay saying it as long as I can.

After we sit down, before I open my mouth, Meghan says, "It's Tay-Sachs, isn't it?" Her eyes wide open, pupils dilated, she doesn't ask, she knows. George's hand claws Meghan's shoulder.

I close my eyes and nod: "Yes."

Meghan gasps and incoherent vowels emanate from her open mouth. George tries to cover it with his hand, but Meghan shakes her head. He hugs her instead, his arms around both his wife and daughter. Meghan squeezes Adèle to her chest and all three rock together.

I want to turn away, to walk out of this office, to run. To not see this raw sorrow and pain. George looks at me, mute, his mouth a thin line, eyes glistening.

"How?" he asks. "Meghan is not Jewish. How?"

It never gets easier, it doesn't stop hurting and I have to do it again.

READING DOSTOEVSKY IN NEW YORK CITY

"Let's have a look at the timeline," Maura said.

I breathed a sigh of relief: I could do that.

We were discussing "A Gentle Creature," a forty-five-page short story by Dostoevsky that had been assigned to us before the weekend workshop and I was quite lost and confused. Eight health care professionals from around North America, eager and interested and intrigued by narrative medicine, we huddled around faux-pine-laminated tables arranged in a circle in a Columbia University classroom on a glorious fall Saturday afternoon in New York City. Outside the Lewisohn Hall, the leaves on the lindens lining Broadway were beginning to turn and the afternoon sun angled through the tall windows. Five physicians, a chaplain, a social worker and a nurse, we all had read the story before coming here and were analyzing it using the tools of close reading. Maura was Dr. Maura Spiegel, a lecturer in the Division of Narrative Medicine at Columbia University.

During this discussion, we were to determine and examine

the story's frame, form, time, plot and desire. A tool of narrative medicine, close reading leads to a deeper understanding of a piece of writing and, in turn, sheds light on the assumptions we make when we communicate in words. Such assumptions and preconceptions may interfere with medical history taking and, as a result, reaching the correct diagnosis. It is an old clinical medicine adage that "diagnosis is eighty percent history": in the vast majority of cases, the correct diagnosis is made by listening to the patient, and physical examination and laboratory tests and X-rays only serve to confirm it.

So far in the workshop, we characterized the story's narrator as direct but pleading and needy and most definitely unreliable, and the narrative as a chaos narrative that occurs at a time of a great upheaval and takes the form of stream of consciousness. The plot was deceptively simple – a man arrives at his apartment minutes after his wife has jumped out the window of their fourth-story apartment, but I was completely at a loss to identify the desire of the story. I couldn't figure out what the narrator wanted, what was driving him, so it was a relief to switch to something I thought would be more manageable: the timeline of the story.

Throughout my medical school studies in the late 1980s, as I was learning anatomy, histology and biochemistry of the human body, I had felt that something was missing. While revelling in learning the intricacies of how the human body was put together – miraculously and complexly – and how it functioned – wondrously controlled and exact – the two things I had always dreamed of studying, I wasn't fully satisfied. I didn't want to be the pure nerd who spewed minutiae from the textbooks at the drop of a hat; I wanted to also be able to discuss the themes of *Anna Karenina* and the poetry of Szymborska. I wanted to be erudite, eloquent, not constrained by the medical school curriculum. On occasion in the teaching clinic, I actually supplemented my differential diagnosis lists with facts I had gleaned from reading literature: cardiomyopathy developing after a childhood bout of diphtheria from a Polish novel I had read as a teenager or the shotty enlargement of lymph nodes in secondary syphilis from nineteenth-century artists' biographies.

As a child, I had always been that cliché – a voracious reader. My studies came easily, so I filled time by reading anything I could put my hands on: world literature classics (my recounting of the engineer of the *Patna*, who saw "pink toads with diamond eyes" while suffering delirium tremens in Joseph Conrad's *Lord Jim*, impressed my clinical skills supervisor), propaganda-filled teenage novels, fables, fairy tales, even – reluctantly – poems but only the ones that rhymed (for many years rhyming remained my definition of poetry). Later, that reading was also an attempt to learn the new society I was thrust into, this Canada I found myself in and didn't really understand. Here, I relied on Robertson Davies and Mordecai Richler. In medical school, I expanded the repertoire to patient memoirs and Oliver Sacks. By then, I was reading to learn about patients' experiences. I wanted to know how it felt to be on the other end of the stethoscope. And when the fourth and final year of medical school came about, the toughest and the most physically and mentally demanding eleven months of clinical clerkship, I relaxed with Wyndham's lichen and chrysalids.

I thought I would be well-prepared for close reading, but I had always been reading as a pastime, for the story and for the good word choices or vivid descriptions. First in Polish where I loved puns and word games in children's poetry, and, after almost a decade of reading and speaking it, finally, in English. But I did not know how to critically analyze a story, hence my distress during the workshop.

*

The goal of narrative medicine can be stated simply – to understand and honour the life stories of persons with illness with the view of improving patient care or, as Rita Charon, a professor of internal medicine at Columbia University and the inventor of narrative medicine, puts it, it is the "ability to recognize, absorb, interpret and act on the stories and plights of others." More and more, medical professionals recognize that illness goes beyond the clinical narrative represented in the medical chart. Ill people speak with a voice that is their own, not that of their doctor, surgeon or counsellor; *they* are the experts on their own illnesses. Narrative medicine teaches clinicians how to hear those stories and voices, to understand them in a nuanced way, to narrate them skillfully and, as a result, to become steadfast in acting on patients' behalf and to provide better medical care.

Narrative medicine attends to the person's situation as it unfolds in time not constrained to the clinic or hospital, but in that person's life as a whole. As practised in the clinical setting, it comprises three distinct but intertwined elements: attention – listening to the patient's story; representation – the retelling of the story to one's colleagues; and affiliation – advocating on the patient's behalf based on her story. Narrative medicine strives to achieve a balance between the patient's story and the doctor's story with the end result of representing them both faithfully and aptly.

And that was what I came to New York City to learn.

Reading "A Gentle Creature" a month before the workshop, I had struggled with the text – it was dense, the vocabulary archaic, the syntax complex. The plot was simpler to analyze as there was not much of one: a man whose young wife jumped out the window of their fourth-story apartment ponders their life together and considers what might have led her to such a terrible act in the immediate aftermath of her suicide. In addition to struggling with the language, I was confused by the unnamed narrator: he alternated between what seemed to be genuine sorrow and a self-righteous indignation at his wife. He never allowed that he might, even partially, be to blame for her suicide. Trying to follow his logic, I became unsure of my own judgment – his insistence at his own innocence stumped me. I wondered what I'd missed; if I misunderstood, misrepresented his attitude: it was Dostoevsky after all, not the easiest read. I backtracked again and again, reread passages. This tactic – so helpful in reading other classic texts – did little to enlighten me: I remained unsure of the narrator's true intentions and motives and, as a result, I began to question my own feelings and interpretation. He couldn't be as awful as he seemed, could he? What was I missing?

In the workshop, we readily agreed that the story was told in first-person point of view, that the narrator was also the protagonist – an easy enough determination; but we argued about his motives, the "desire" of the story. We outdid each other in showing how horrid he had been, how repugnant. We agreed that his actions were reprehensible, but we couldn't agree if he was a narcissistic scoundrel who manipulated and psychologically abused his wife or only a well-intentioned blunderer. I felt relieved that I wasn't alone in my confusion. The group also seemed stuck about the motives and I was none the wiser.

Maura's redirection to concentrate on the timeline was a welcome change. But a timeline analysis proved difficult even with a graph drawn in chalk on the blackboard. The text backtracks and meanders, with flashbacks and flash-forwards and asides. The narrator contradicts himself regularly and tells many side stories not relevant to the main plot. The graph on the blackboard looped back, split apart, came together, split again and still I had no better understanding of what we were supposed to get out of reading this story.

*

In narrative medicine's close reading, the initial task of reading a text is the identification of the story's narrator. The same holds for clinical encounters where the narrator may not be the obvious one: Is it the mother of the baby identified in pregnancy as having brittle bone disease and multiple fractures, or is it the fetus itself? Is the voice that of the patient's mother who has Huntington disease or those of the affected woman's potential grandchildren? Is the protagonist of the clinical encounter the boy affected with a deadly inherited disease or the parents who have just learned that they are carriers of a lethal condition?

A physician attuned to what constitutes the frame of a story will be more likely to see what is missing and ask for overlooked symptoms; a clinician who is more aware of the timeline will detect gaps or correlations between illness episodes or whether we are dealing with an acute or a chronic illness. Ability to recognize plot twists and turns and nested plots will aid in identifying the different stories of different narrators affected by the illness – a crucial piece of information while gathering family history and during genetic counselling.

As a clinical geneticist, I hear stories all the time: accounts of pregnancies, miscarriages, deliveries, life stories of children born with multiple birth defects or dying with progressive diseases, of family members who are affected by the patient's disease or who have disowned her. The narrative medicine approach to medical care arose in response to the progressive dehumanization of medicine over the last three decades. Internal medicine has become a specialty where computer printouts have replaced handwritten notes, where doctors look at monitors instead of checking a patient's pulse and never measure blood pressure themselves. I trained in the mid-eighties when the most important and validating purchase on the first day of medical school was a Littmann stethoscope and an ear-nose-and-throat examination kit: otoscope and ophthalmo-scope from Welch Allyn. We spent hours learning to listen to heart sounds and visualizing ear and nasal cavity landmarks, first on our-selves, and later on patients. Listening skills are now forgotten, no longer valued. We had a six-month course in communication skills where we trained to interview patients, to listen to them, to notice them, to observe and gather information from appearance. Maybe my classmates who later specialized in internal medicine or surgery forgot those skills so that by the mid-nineties it was all technology and machines. In contrast, genetics is a very loquacious specialty. Geneticists, and especially genetic counsellors, spend hours talking to people: explaining genetic concepts in layman's terms takes skill and time.

As a medical student, I memorized graphs of the respiratory cycle of the cell and of the electrical discharges during the heart muscle contractions, diagrams of intracellular biochemical reactions and of DNA synthesis. I studied the chemistry of kidney function where numbers reigned supreme. I traced with my index finger the electrical discharges of the human brain recorded during sleep and during seizures. But soon I realized that people as a whole didn't follow rules that apply to their parts, they didn't read the same set of instructions as doctors did, that most of clinical medicine was an exception to the rules we learned. Over decades of practice, the communication skills I learned in that course in the first year of medical school allowed for the necessary flexibility in my clinical approaches. With narrative medicine training, I gained even more insights. An appreciation for silences. A patience for convoluted, poorly organized medical histories, which I (subconsciously by then) analyzed with the narrative lens to gain insights I would not have had without its benefit. An instinctive recognition of speech patterns, dialects and accents. I was building on years of previous clinical experience but after the workshop the insights came more easily and were welcome as clarifications and not as unnecessary interruptions or intrusions into the patient's discourse. Letting a patient speak provided not only medical information but also, not surprisingly, a therapeutic effect of having been heard.

In the workshop, an internist from Long Island finally said, "He is such a poor historian." He sounded exasperated. "I can't tell what is true and what isn't."

A poor historian. How many times have I heard that phrase used in our teaching clinics? It refers to a patient who rambles on, insists on telling what appear to be unnecessary details, deflects all attempts at redirection with yet another seemingly irrelevant anecdote. These patients are the banes of medical students' existence because timid and new as the students are at taking history they can't get to the crux of the matter, and of busy clinicians who have dozens more patients in their waiting rooms or emergency departments and cannot afford the time. And while a more seasoned physician may be able to redirect and ask close-ended questions, such an interrogation often leaves both sides unsatisfied.

Narrative medicine invites patients to tell their own stories to their physicians, to other patients and to society at large. Narrative medicine views patients as texts to be read, pondered and respected. Hippocrates himself said: "It is better to know what kind of a man has a disease, than what kind of disease has a man." Until the ascent of modern medicine to its logico-scientific apogee, shamans, medicine men, wise women and physicians treated illness as part of life and attended to the whole person, his family and his community. When they could not treat the physical ailments, they listened: listening to and telling of stories were part of treatment.

In 1905, at the beginning of the modern medical era, the Canadian-American physician William Osler admonished his students to spend "the most important half-hour of the day" reading literary works. In 1992, Anatole Broyard, in his poignant and often ironic memoir of living with cancer, said: "I don't see a reason why doctors shouldn't read a little poetry as part of their training." Reading literary texts we vicariously experience the fear and pain and anger of individuals with life-altering diagnoses, of people living with chronic illness or of those nearing death – their own, their child's or their relative's. Literary masterpieces frequently are centred on illness and its sufferers, on death and our desire to understand it, and as such they provide a rich and fertile substrate for empathy.

In the end, reading "A Gentle Creature" was an exercise in ascertaining the history of the "patient," in this case, the narrator. The story trained us to listen to the "hateful" and "difficult" patient and to rise above those feelings for the patient's sake. Teasing out the timeline and the frame and the desire of the narrator helped us to "get the history" – to understand better what he might be trying to say or to withhold. And my confusion and doubt about his motives was a good example of countertransference, which is defined as the physician's emotional response to the story that the patient is telling. Our group reading of "A Gentle Creature" showed me how close reading hones physicians' interviewing skills. By questioning the plot, the identity of the narrator, his desire and the timeline of this complex story we were training ourselves to listen better and to hear the stories that were being told to us. All for the patients' benefit, no matter how difficult or "poor" a historian they are.

All these years reading Tolstoy and Atwood and Reymont I have been training myself to be a good doctor.

There is no better example of the utility of narrative medicine than that of taking care of a patient with hypermobility subtype of Ehlers-Danlos syndrome. It is a connective tissue disorder that results in lax ligaments and loose supportive tissues of joints and internal organs. People with this condition have myriad complaints that frequently cannot be explained with classical approaches to clinical medicine. Chronic abdominal and pelvic pain, muscle and joint aches, joint dislocations and laxity, feeling of "looseness" and pain in the abdomen, elastic skin that bruises easily, labile blood pressure that causes fainting and lightheadedness, rapid heart rate that may cause chest pain – all these may lead to secondary psychological issues such as depression and anxiety, chronic disability, job loss and difficulties with interpersonal relationships. There is no treatment for hypermobility Ehlers-Danlos syndrome other than management of its complications; there isn't even a genetic test for this subtype. "All tests are normal" is what these patients frequently hear, but their lives are anything but normal. I have nothing to offer these patients other than to sit and listen to their litany of symptoms. I let them tell the story that many doctors have interrupted or, worse, disbelieved. So I listen and acknowledge their pain and discomfort and suffering. It takes time, but time is the only thing that I can give them.

I think my patients like that.

METANOIAS: ON NATURE AND LANGUAGES

"The bluebells had not faded yet, they made a solid carpet in the woods above the valley, and the young bracken was shooting up, curling and green." So read a sentence in the middle of du Maurier's *Rebecca*, as the unnamed narrator walks through the Happy Valley to the cove. "Bracken" had made an appearance several times earlier in the pages of the novel, yet it is only now that the verb "curling" evokes an instant recognition in me: fern. Fern was bracken? The English word echoes "brackets" and "brackish"; how do those spiky and rectangular and sea-salty words connect with the fern? A plant I have loved as the underbelly of my childhood forests, the magical *paproć*? That held me in thrall because it blossomed only on Midsummer night and brought luck to those who found the bloom, as Polish folktales of my childhood foretold?

A plant I have known all my life, with an English name I thought I knew, is suddenly called something else. My world shifts on its axis, just a bit.

How do you do, bracken. Nice to meet you.

*

"Mushroom" was easy to remember when I was learning English because it reminded me of *muchomor*, the Polish word for the red-capped, white-spotted amanita of the deep forests and fables. Deadly, even if it does serve as the house for forest gnomes, the danger only adding to its mystery and appeal. Once, foraging for blackberries in the forest behind my primary school, crouched among the brambles, I watched a fly crawl on the *muchomor*'s scarlet dome, seemingly immune to its toxin. I tapped the cap and the fly flew away, buzzing. I saved it, but was my fingertip now poisonous?

Had I known, the English name would have taught me right away that the fly was doomed. Does "flybane" sound as ominous to English ears as *muchomor* does to mine? Does "death cap"? I doubt that boletus or king mushroom has the same amazing associations with culinary delights as the Polish *borowik*, a word that brings to my mind its unmistakable aroma and the flavours of hunter's stew and Christmas mushroom dumplings.

The summer evenings of my Polish childhood were scented with Matthias flower – *maciejka* – a profusion of tiny, long-petalled flowers in pink, white or purple. By June, the fragrance floated through our open windows from the garden of one of the neigbouring houses dwarfed by our five-story concrete apartment building.

I forget about it until twenty years later, when one summer evening, sitting on the deck in my own garden, I realize something is missing. The scent. From the recesses of my olfactory memory, *maciejka* wafts into my consciousness. I wonder what it is called in English and whether it would grow in southern Ontario. The name found in a gardening manual – evening stock – disappoints. Its Latin name, *Matthiola*, honouring Pietro Andrea Mattioli, an Italian Renaissance physician and botanist who first described the plant, echoes the Polish, but the English is dry, scentless.

I still haven't planted it. Just like its English name, I'm afraid the poor little flowers and their scent will fail to live up to my memories. It's easier to leave it in the past than to face disappointment. Maybe if I ordered the seeds from Poland, it would be different.

A brown streak alights on my bird feeder and resolves itself into a rufous-crowned sparrow. Thanks to the Audubon Guide to North American Birds, I also greet a white-breasted nuthatch and a tufted titmouse. Naming the species gives me a sense of accomplishment, but it is only when I look up their Polish translations that a wide grin breaks out on my face: *kowalik* and *sikora*. And *kowalik* – a little smith – is such an appropriate name for this smaller distant cousin of woodpeckers. These two – along with many other birds – populate traditional poems and songs, occasionally flitting on the winter windowsills of my childhood. Only after I read their Polish names do I feel that I know them, these birds from Jan Brzechwa's children's poems – although I notice the American titmouse does not have the canary-yellow belly and the jet-black cap sported by the Polish *sikora*.

The two books from which I have learned the most about gardens and nature were, ironically, the Polish translations of the *The Secret Garden* and *Anne of Green Gables*. Anne continually goes into raptures about the avenue of flowering crabapple trees or white narcissi in the forest behind Matthew and Marilla's house or heather in Kingsport's Prospect Point. With only a frail little tree planted to soften the concrete wasteland outside my window as my companion, I drank in the nature between these pages. And the moors of the Polish translation of *Jane Eyre* were so evocative because they were translated as *wrzosowiska* – heather fields. They might not have evoked the same dread as "moors" do, but they allowed my child's imagination to roam amid the heather and the shrubs and the windblown, stunted trees.

In grade eight Polish, I learned that *poziomka* – wild strawberry – was so named because its tendrils stretched *po ziemi* – over the ground. Was strawberry called so because it grew on the straw of last year's grass? Buttercup is called *kaczeniec* – duck flower – because of its brilliant duckling yellow, but why duckweed? The Polish *rzęsa wodna* – water eyelash – is so much prettier. Other English names are self-evident: lily of the valley grows in ground depressions, tiger lily for its vivid colour. But even if I understand the etymology of the English nature names, only the Polish resonate in the centre of my being. The difference between feeling and knowing, living and learning.

Memories live in the deepest grey matter nuclei in the brain, the amygdala (named so because it looks like an almond) and the hippocampus (shaped like a sea horse). These two structures belong to the limbic system, the most ancient, most primitive part of the brain, the seat of instincts and impulses and hunches, the one we have in common with reptiles. The olfactory lobes connect to the limbic system without processing through the higher cortex, and that's why scents and smells elicit memories so strongly, even the most deeply buried and hidden ones. Like *maciejka* returning to me on a summer evening after more than twenty years.

"A rose by any other name would smell as sweet" claims the Bard. I beg to differ. The names we learn in childhood, the ones that grow with us and with use, that acquire layers of meaning, smell the sweetest to us. They hold memories and stories in a way that a name studied, learned later in life does not; people who don't share your experiences or language or culture may not fully appreciate those associations and feelings. They have their own associations, their own meanings curled inside those common nouns. It can be like sharing a secret, or a sort of magic, the act of translating them for one another. Is there a word for this kind of intimacy?

And what of the bracken? It turns out that it is not the *paproć* of my childhood or the shady recesses of my Ontario garden slope after all, but of a particular species that grows in England and Scotland. I'm surprised to learn that in Polish, the plant is named *orlica* – the term for a female eagle. It is so named because of the evocative shape of the unfurling frond in the springtime. First clutched closed, then partly open, it suggests the movement of an eagle's claw. *Orlica* is a strking word, rough and powerful, with nothing to suggest the soft curved fronds of the *paproć*; it belongs on rocky outcrops where only eagles dare to nest.

The soaring maple tree in my garden – strong and limber enough to withstand the derecho thunderstorm that felled five trees around it last year – is footed in a sea of ferns. These are the ferns of my childhood, gracefully lifting their fronds from the earth. Here, in July, I watch fireflies dance on currents of warm air as they send their lovestruck light signals. I have never seen fireflies in Poland, even though I knew their name – *świetliki*. They belong to the forty years of memories of my summers on Georgian Bay and in my Hamilton garden, to the new memories I have made with my husband and sons.

SENSORIUM

Sensorium: originally: a supposed seat or centre within the brain in which sensations are united (now historical); (also) a sense organ (obsolete). In later use also: the sensory elements of the nervous system collectively.
– *Oxford English Dictionary*

Smell

The fragrance catches me mid-step; a potent floral, it tickles the back of my nose and throat. Honeyed, ambrosial, sweet – almost *too* sweet. In the middle of the University of Toronto's Victoria College quadrangle, I am transported three thousand miles east and three years earlier, back to Poland. Sniffing the breeze, I spin around and around to locate the source. There! An unimposing tree in a stone-studded concrete planter about thirty metres away, its branches barely reaching the second story of the Northrop Frye Hall. As I near it, I recognize the heart-shaped leaves and the deeply grooved, greyish bark on the slender trunk I can circle with my hands. It's a . . . What *is* it called in English?

"What's the name of this tree?" I ask a passing student.

He goggles at me, then he shrugs. "What am I, a forestry student?"

How can you not know the name of the trees that grow around you? I wonder if all Canadians are like that. I know it as *lipa*, pronounced LEE-pah; in Polish, its name shares the origin of the name for July – *lipiec*. Its domed crowns tower over fields and line country roads, palace driveways and city avenues; a staple of my childhood landscape – a landscape where I no longer belong and which I suddenly miss with my whole being.

It will take a while to discover the tree's English name – it was the summer of 1983, decades before instant internet searches or translations. The four-volume Polish–English dictionary I lugged across the Atlantic when my family emigrated suggested "lime" as a translation, which brought to mind the sour green citrus and not the trees of my childhood. It wasn't until I read Austen's *Emma* that I understood that "common lime" *was* the English name of this tree and that, just like in Poland, it grew in espaliers along manor driveways. In Europe, it is also known as linden (from German, with *Unter den Linden* – "under the linden trees" – being the throughway in Berlin that leads from the Brandenburg Gate to Kaiser Palace) and, in America, as basswood.

Had I known its Latin name – *Tilia x europaea* – I probably would have been able to learn much sooner what the tree's English name was, even if what I was smelling was *Tilia x americana*, the linden's North American cousin.

I had learned most of my trees from reading about them as a child, not from seeing them. Many I knew only from description:

weeping willows dip their long branches into village streams, slender alders whisper on the mountain meadows, palms sough and droop their fronds. Several types of conifers appeared in our apartment at various Christmases: balsam fir was the best because its needles lasted the longest while the pine's were too long to hook ornaments. Three species I knew better – oak and horse chestnut because of the nuts and acorns they dropped every fall and the unmistakable black-and-white-barked birches. But *lipas* I knew with all my senses.

Proprioception

A thick branch straddled between my ten-year-old thighs, hidden in the deep emerald shade of the cordate leaves and sky-reaching boughs, the leaves murmuring in the faint breeze. I have climbed the century-old linden tree growing in my aunt Adela's yard to pick its blossoms. I'm dizzy with height and the overpowering scent. My ears buzz with the bees – thousands of them descend on the tree at dawn and work away until dusk, collecting pollen. Droplets of clear nectar shine on the tiny flowers, sticky on my fingertips and palms. Minuscule white petals, green stamens dotted with golden specks, each cluster protected by a pale green bract, the whole shorter than my pinky finger. The aroma is like a living thing wrapping me in its folds. I inhale, deep, deep, and almost lose my balance.

Hours spent climbing from branch to branch, filling my basket with thousands of bracted flowers and bracts, sliding down the trunk only to empty it and clamber up again. From the tallest bough, where the shade is dappled, I loom over the village's red-tiled roofs, even over the weathered wooden slats of the barn. The pine forest on the horizon is pierced by the church steeple five kilometres away and the chimes of the bells reach my ears as they toll for noon mass. Black-and-white cows dot the meadows, a horse drags a wagon back from the farmers' market in the city, wheat fields ripen gold behind my aunt's house. This is what flying must be like.

"Małgosia! You'll fall!" my aunt calls from below, her neck extended as she struggles to see me, her hand on the thick bark of the tree for balance. "Come down – we've got enough flowers to last three winters."

I linger on the bough, my feet dangling. I imagine my aunt spreading the blossoms on a baking sheet where they will dry in the heat rising from her charcoal-burning stove. When dry, she will transfer them into glass jars where they will await my cousins' first autumn sniffles and coughs. She will scoop two heaping teaspoons of the dried flowers into a mug, pour boiling water over it and let them steep for five minutes, steaming, the pale lemon-yellow linden herbal tea a remedy for colds and fevers. And a memory evoker, for it was his aunt's linden-flower tea that Proust recalled upon tasting the madeleine that fateful afternoon.

Taste

Decades later, in the warm kitchen of my own home in Hamilton, I lift a stainless-steel spoon out of the honey jar – the thick pale-yellow liquid coats the spoon's bowl and flows down in a wide ribbon that folds on the surface before it sinks into the honey with a tail of tiny bubbles. A woodlands' scent with hints of menthol, camphor and mint – a noseful. When fresh and raw, linden honey is a clear liquid with a hint of green; it darkens to amber-gold when mature and is studded with sugar crystals.

Linden-flower honey is produced by bees in hives set in proximity of a linden tree. A cure for the common cold and stomach flu – often doubled up with linden-flower tea. Avicenna, a medieval Muslim philosopher and physician, claimed that he had cured epilepsy, headaches, rheumatism and kidney diseases with linden honey and called it liquid gold. By volume, forty percent of honey is glucose and because of this, pure honey is quick to crystallize – only non-artificial, non-modified honey extrudes yellowed sugar lumps onto its surface, giving the honey a granular texture. Organic apiarists extol its electrolyte and microelement composition for its resemblance to human plasma, but it's a pure coincidence, of no medicinal value. Honey contains vitamin B1 and B2, niacin, vitamin C, biotin and vitamin K but not in significant amounts – one would grow quite fat if honey was the only source of those micronutrients in the diet. It is also rich in tannins which give honey its astringent properties.

After licking the spoon – the honey sweet and delicately sour, with a trace of a bitter aftertaste – I scoop a spoonful and pour it from a height into a mug of steaming Assam tea. I watch it curl and dissolve into the amber liquid. Honeyed tea – the best drink for reading on an early December evening, as dusk falls and the windowpanes glow like mirrors, my living room doubled, hazy, beyond the glass. It would be even better if it were snowing I think as I curl on the sofa under a wool blanket, a fragrant cloud rising from the tea on the coffee table, a book in my lap.

Hearing

The small ebony disc begins to rotate, the needle drops and faint scratches fill the room. *Plink-plink-plink* – three short, happy notes, each higher than the previous one; a lively waltz. A violin pizzicato followed by a boyish tenor singing in Polish:

> *You shall gather flowers*
> *You shall smile a lot*
> *You shall count the stars*
> *You shall wait for me*

And you, just you, shall be my lady
And you, only you, shall be my queen.

My feet in their small square-nosed sandals follow the rhythm and I glide across the rug, dipping with every third step. I twirl, bow to the "lady" and to the "queen," imagine myself in a medieval castle with a golden cone of a hat draped with a veil that follows me, fluttering, as I stumble on my pirouette.

The song came out in 1967, so I might have been four or five years old when the 45-record made its way to my parents' mono turntable. Styled on medieval troubadour's songs, it promised the object of affection linden fiddles playing and rowan groves singing just for her. I still think that "You Shall Be My Lady" is the loveliest love song of all time and its ending "you shall have love / like the autumn storms" so erotically charged that even my child's mind thrilled with it.

Fiddles carved from linden wood have the sweetest sound, or so the lore goes; linden wood's fine, even grain, pale white or light brown, devoid of knots and gnarls, lends itself to easy shaping. Folk fiddles were created differently from standard violins: the body was sculpted from one piece of linden wood, rather than from separate blocks for the back and the ribs, and topped with a spruce wood plank. Sometimes even the neck was carved together with the body. The bridge stood on uneven feet with the side that supported the D and G strings taller and resting on the body's bottom while the other side rested on the top.

Linden fiddle sounds higher in register than violin, sometimes squeaky, sometimes tinkly. It warbles happily in dance music, thrums in your chest. The voice coaxed out of nature, the music of the tree released into the air, singing, singing high up into the sky where it tangles in the linden boughs reaching for the cloud. Village bands played the fiddles at harvest festivals together with a *basetla* – a string instrument almost as large as a cello, with a deeper sound that resonated in your bones. Together, they lifted the soul above the village roofs, above the church steeple, into the air and – however momentary in the life of an indentured peasant – happiness.

Touch

On my walk home eighteen years after my encounter in the Vic Quad, I noticed the four little linden trees only the second summer we lived in our new house – we had moved in August, well past their blossoming. I had not visited Poland for all those years – trying to establish a life in Canada, I was reluctant to return and face the life I could have had in Poland.

Until that day, I hadn't recognized the scrawny trees strangled by paving stones in front of McMaster University's John Hodgins Engineering Building even though I had marched past them twice a day. But that hot and humid July their fragrance enveloped me – by then I had learned to recognize the tiniest molecules of the scent and could trace them like a hound dog. I gazed at each of them in turn, thrilled that they would guide my walks home from work. But farther up the street, already in our division, a much larger surprise awaited – a mature linden tree threw a forties bungalow into a deep shade, its branches abuzz with honeybees and bumblebees. It stood rooted in a small rectangular lawn, the tallest tree by far on this street corner. Suddenly not in a rush, I ambled toward it and lay my hand on the trunk wide enough to be circled by two people's arms, the grooved grey bark radiating the warmth of the day. Looking up, I saw that it was taller and wider than Aunt Adela's tree and I remembered climbing and riding similar thick, gnarled branches.

"Hello?" I heard a question in the greeting.

I turned around, my hand still on the tree. A middle-aged woman peered at me from the porch.

"It's my favourite tree," I said. "Sorry."

"It's a linden," she said, her voice bearing a trace of an accent I couldn't place.

I smiled as I walked away. Finally, a Canadian who knew her trees.

Except that she wasn't. I met her again at a fundraiser for the chair of Polish language at McMaster University the following February. Nina was Polish, a Slavic languages professor at McMaster University. Educated in Warsaw, she had met her Russian-Canadian husband in Moscow where she had been studying Russian on a student exchange in the sixties. We chatted about how we had met under her linden and she told me that when she and her husband bought the bungalow in 1978, the linden was already towering over it. She thought that it might have been planted when the house was built sometime in the late forties, making it at least eighty years old, long before either of us was born or arrived in Canada.

At that Proustian moment in Vic Quad forty years ago, the smell of linden blossoms revealed my homesickness and isolation in a new country. Over the years, I have made peace with my memories and my dreams, and now Nina's linden bestows a sliver of Poland upon my Canadian existence.

Vision

In my basement library, looking for a book Nina wanted to borrow, I scan the shelf with my old Polish books, the ones that travelled with me across the Atlantic. A three-volume embossed linen set of Jan Kochanowski's writings stands among them. Also known as the Bard of Blackwood, this sixteenth-century poet was the first to use the Polish vernacular instead of Latin in his poetry and dramas. He is most famous for his tragic *Laments*, written after the death of his precocious three-year-old daughter Ursula, and for his trifles, short occasional poems about everyday objects and happenings.

I pry Volume One from the tightly packed shelf. The small hardback fits lightly into my hand. I check the index, my fingertip tracing the rough pages beiged with age, and leaf to page 181.

"*Gościu, siądź pod mym liściem, a odpoczni sobie . . .*" I read.

Traveller, come. An exhortation to rest under its leaves begins "On the Linden Tree," a short poem first published in 1584. Set in the Polish alexandrine: a soothing rhythm of thirteen-syllable metre with a caesura after the seventh syllable, this trifle is twelve lines long. My back against the fake wood of my bookshelves, I read about a living and giving tree. My eyes follow the printed lines and I hear the Polish words clearly in my mind as if I were reading aloud.

It's all there: the "cool shade" of the graceful boughs, a shelter for humans and birds alike; the buzzy bees gathering nectar from the "sweet-smelling flowers"; the honey that "graces the finest of tables"; the soft soughing of the leaves in the summer breeze that "sing [. . .] visitors" to sleep (translated by William Auld). My sight, tracing the letters on the page, transmutes into other senses as if by magic – I hear the wind, I smell the blossoms, I taste the honey, my palms twitch against the grooved bark.

Poetry's magic. I am transported. All my experiences and knowledge of the linden resonate, echo in my brain and coalesce into an ur-experience of the linden until I feel – I *know* – the linden in all its leafy, barky, blossomy linden-ness.

THE ART OF SEEING

"It's an embrace."

"What do you see that makes you say that?"

"There are two figures, one on top of the other, their limbs are intertwined. The upper one is clearly a male, but I can't see the genitals of the other."

"Anybody else?"

"The arrangement of limbs in the lower right corner really bothers me. It suggests pain, a lot of pain, to me."

"What do you see that makes you say that?"

"The knee is bent forward, not anatomically, this suggests to me a complete destruction of the knee joint, which must be very painful."

"Anybody else?"

"What-if this is an aerial view, the entire sculpture is a bed and these are two figures lying down side by side, post-coital and not one dominating over the other?"

"There are too many fingers and toes on some of the limbs, and not enough on others, but overall it adds up to ten, so it is complete."

Twelve people huddle in a gallery at Boston's Museum of Fine Arts on a fine Sunday morning in October. They are gathered in front of a plaster bas-relief hung in the corner; the deep, rich maroon of the wall a perfect counterpoint to the ivory of the piece. In the corridors, sun floods the floors but here the artwork is bathed in light from bright ceiling spotlights. We are to offer our interpretations of the work in front of us to the questioning instructor. Uncertain, we muddle through, our perceptions as diverse as we are: physicians, social workers, a physiotherapist, a chaplain, a psychiatrist. A medical humanities professor. A geneticist.

Our comments reflect our training, our life journeys and our experiences – no two are the same. In a group of twelve people, there are twelve distinct ways of seeing the same piece of art.

At the end of the discussion, I ask for the title of the artwork – it is *Wrestlers* by Henri Gaudier-Brzeska, from 1914. In an instant, the whole picture reshapes itself completely from the embrace and love I saw in the arrangement of the figures and clicks into a different understanding. But which is true? Which one is – for the lack of a better word – correct? Love or struggle? Conflict or erotica? Both? Neither? An unsolvable conundrum.

I am at a workshop held by the Centre for Narrative Practice and by Arts Practica, a medical consulting group run by the painter-turned-medical-educator Alexa Miller. It is titled "Aesthetic Attention: Art and Clinical Skills." By decoding health care professionals' approaches to visual arts, it is intended to help them see things better, to be more attentive to details of patients' clinical presentations and to identify the biases that colour their perceptions in a clinical situations. All with the ultimate goal of improving patient care. We spent the first day in a classroom where Alexa led us through a number of exercises, looking at slides of paintings, art photographs, sculptures, and asking us what we saw. She didn't ask for reflections – only for what we observed – then challenged those with a gentle yet probing: "What do you see that makes you say that?" over and over until we articulated what it was exactly that made us see what we thought we saw. This lengthy process also allowed us to hear others' statements enacting another goal of the workshop: to communicate with respect for the perspective of others.

Between the exercises in art interpretation, an internist who is a clinical skills educator at Harvard Medical School reviewed for us the science behind bias and cognitive error in medical diagnosis. From her talk about fast and slow thinking, I realize that, as a clinical geneticist, I frequently employ "system one thinking" – the spot diagnosis, the *Augenblick* of instantaneous interpretation of features and signs. System one thinking is our brain's automatic, unconscious response to situations and stimuli such as absent-mindedly reading text on a billboard, tying our shoelaces without a thought or instinctively hopping over a puddle on the sidewalk. In my case, the system one response to a patient with short stature, wide neck, heart defect and characteristic facial features leads to the diagnosis of Noonan syndrome or to thinking "achondroplasia" before naming the actor Peter Dinklage. This pattern-recognition paradigm serves my specialty well, but this type of thinking is why, anecdotally, the most commonly missed fracture in the emergency room is the second fracture: the clinician focuses on the bone that the patient complains about, casts the broken limb and does not look any further. That is also why highly trained radiologists asked to look for cancerous lesions of the lung miss the tiny, literal, black-and-white gorilla camouflaged on a chest CT. I missed the gorilla as well when Alexa showed the image to our group, proving that I, too, suffer from inattentional blindness: seeing only what I am looking for.

On the other hand, the exercise of looking at art and describing what we see helps us slow down and engage in system two thinking – the deliberate, slow analysis of available data. Proceeding consciously, we don't miss clinical signs and symptoms hiding in plain view; this slowing prevents erroneous jumping to conclusions. I am here to relearn how to notice and to see, to reacquire the reasoning skills I was taught long ago in medical school: to articulate thoughts in an orderly and attentive manner.

The next day, we used the slow approach to examine other works at the museum, including a see-through, half-unravelled portait of a young man made in – astonishingly – macramé and an enormous painting that filled the wall of the exhibit room with white silhouettes on primary colour background depicting the brutality and dehumanization of slavery. In both cases, my fellow workshop participants offered varied reflections, sometimes agreeing with mine, sometimes diametrically different but always thoughtfully led by Alexa. Afterwards, we stepped from the darkened halls of the museum into the bright sun wiser and gentler in our responses to others' points of view.

During the two-day workshop, I became conversant with the different ways of seeing, able to recognize biases in perceptions and mindful of not jumping to conclusions. I appreciate that there is no "right" way to see a work of art because art itself – and discussion of art – defies certainty. That, when interpreting art, there is no way to be wrong and that the title of the work should not limit our perception but serve as an invitation to see it more fully, just like a patient's diagnosis should not limit our understanding of her wholeness, but be a means of entry to understanding her health and life journey. Rapid clinical diagnosis has its place in my genetics practice, but a systematic analysis of all the presenting features will only strengthen my diagnostic acumen and accuracy. I learn to slow down and pay attention, to evaluate and augment the system one impressions by system two analysis.

That evening, as my plane banks north after taking off from Boston Logan International Airport, the lights of Back Bay glimmering in the night, I know that I will take Alexa's gentle voice asking, "What do you see that makes you say that?" to outpatient clinics and hospital wards with me for years to come.

WHAT WAS MISSING

"Show me how you climb," I said to the little girl in my clinic room and patted the examining table. Sophie[1] was only the second patient I was seeing in person in our outpatient department since the beginning of the pandemic. It was the middle of August 2020, and our clinic – pediatric genetics and deemed non-essential – had been shut down since March 17. But this four-year-old girl needed abdominal ultrasounds and blood tests every three months; twice a year, she would come for a physical exam. All this a part of tumour surveillance for Beckwith-Wiedemann syndrome, a genetic condition that carries with it an increased incidence of intra-abdominal cancers. As is typical, Sophie's identical twin sister did not have the syndrome, it was caused by an imprinting error that occurs during the twinning of a single fertilized egg.

The second-best way to make friends with a pediatric patient is to ask them to do something they are not allowed to do at home,

---

1 Name changed.

like climb an examining table. The best? To ask them to jump while standing on the examining table, holding their hand, of course – a good neurological test for balance and coordination, in fact. As is climbing.

My little patient scaled the table with a hand from her mother and lay down on her back, her inquisitive little face intent on my masked one, her big blue eyes watching my every move. I lifted her tiny frilled empire dress, yellow with tiny white flowers, and pressed on her belly, felt for masses, the Little Mermaid smiling at me from the front of her white undies.

I touched my hand to the warm skin of her tiny belly. Her abdomen was soft, there was no tenderness on pressing down on her liver and spleen. "Everything's fine," I said and smoothed down the skirt of her dress, laid my hand flat on top. "We're done."

She stared at me and said, her voice sweet with that intense sincerity of a child, "Thank you, Doctor."

I inhaled audibly. There was only so much that I could do not to hug her. Those unprompted words, so earnest, so real, brought me close to tears.

I had missed being with patients so much.

The sister tugged at my skirt. "Me too," she said, not wanting to be left out, and, of course, I obliged. Sophie scooted away on the crinkling paper, and the twin climbed next to her. I lifted the skirt of her identical yellow dress and pressed on her soft, warm abdomen. "Your tummy's fine, too," I said. She nodded sagely and slid off the table.

As a pediatric clinical geneticist, I have not seen a single patient with Covid-19 during the entire pandemic. In March 2020, the disease spread out of China and wreaked havoc in Italy, the media choking with news of intensive care units and emergency departments overflowing with patients with pneumonitis and respiratory failure, of thousands of deaths among the elderly and the frail in Lombardy and Liguria. And Hamilton's McMaster Children's Hospital, where I work, was preparing for the worst. I – all of us – had no idea what to expect. Bodies in refrigerated trucks in hospital parking lots like in New York City? Terrified octogenarians dying in hospital hallways denied the sight of familiar faces, the touch of their son's or daughter's or grandchildren's hands? Respirator shortages? Our hospital administration circulated an email questionnaire to all the physicians about what we would feel comfortable doing if we had to be redeployed. Redeployed? Like in the army? My mind reeling with possibilities, each more dire, I ticked the box next to "Screening." I wasn't being flippant: the other choices – emergency room triage, hospital ward attending or intensive care physician – were beyond my professional reach. I no longer possess the skills to work in the emergency room or the intensive care unit; even a general pediatric ward would be a stretch. I wasn't afraid of Covid-19: I believed that with proper precautions and infection control I, a healthy fifty-six-year-old, would be safe. Asking people about their symptoms and contacts and travel history was the only thing I felt qualified to do. I have not started an intravenous line or prescribed medication since 1997 – for twenty-five years I have devoted my career solely to diagnosing genetic conditions in children and fetuses and to genetic counselling of adults. How could I care for critically sick patients when all I can do is examine bodies for genetically significant minor anomalies? My skill set now consists of identifying and naming those telltale features and cohering them into recognizable syndromes, and ordering and interpreting the results of genetic tests. In the clinic, most of what I do is talk: explain complex genetic concepts in layman's terms, inform the parents of the dire prognosis of their child's condition or consent patients to more involved genetic testing.

*

When lockdown was announced in Ontario on March 13, 2020, all outpatient visits were cancelled. Our clinic manager began the task of reorganizing the clinic to see patients by video. The administrative staff, after having to cancel en masse all of our scheduled patients, were being certified to use the Ontario Telehealth Network and learning how to book online appointments. I was uneasy. I had never seen a patient that was not in the same physical space: so much is communicated by the way a woman tilts her head and cups her hand over her pregnant belly, by the way the father brushes the hair off his daughter's forehead, by the parents sitting together or apart. The technical computer aspects were only a part of my worry. How would I tell a pregnant mother via a computer screen that her child had trisomy 18, a chromosomal anomaly that would render him mute and unresponsive and intellectually disabled? Or, worse, that we had no answer for the constellation of troubling black-and-white findings on the prenatal ultrasound? In the clinic, I reached out to touch my patients when they were upset, offered hugs if they seemed appropriate, or sat in silence, but would my care and concern translate across the video screen?

I had to consult on my first pandemic prenatal case even before I was certified to practise telemedicine, so the counselling had to happen by phone with the parents listening to me on their speakerphone. I spoke to them from my back deck on a hot May afternoon, the blossoming lilacs of my garden scenting the air, so unlike the sterile atmosphere of the clinic room. I couldn't see the mother, I could only hear her voice – anxious and tense – and imagine how upset she must be. And even though it was a straightforward case, likely a normal baby with three tiny cysts studding the inner lining of the brain, I paced the deck like a caged lynx. After I reported the normal chromosomal results to the parents – trisomy 18 had been number one on the differential – they said they would continue the pregnancy. The whole time I felt like I was talking in the dark. I missed seeing the mother's eyes and face, seeing how she reacted in her body: tense shoulders? Slumped back? All I had to go on was her intonations and the words she chose. "I wish I could have been there with you," I said at the end of the call.

After being certified for telemedicine, I consulted on the next prenatal case by video: a fetus who initially was thought to have short limbs and a small chin. The mother had had an amniocentesis and this baby's chromosomes were also normal and, with the ultrasound images now showing normal growth, I could reassure the mother that things appeared to be on course. And it helped to see her without a mask covering the bottom of her face and to notice that she, too, had a very small chin.

"Yes," she said when I mentioned it. "Until now, nobody has seen me without the mask." Not being able to assess the faces of the parents to determine if the baby's features were pathologic or familial became yet another casualty of the pandemic.

On another call, a couple booked for counselling about the recurrence risks of tuberous sclerosis kept their appointment only to tell us they'd had a miscarriage the day before. They were both crying and there was nothing we could do but offer our condolences. In such a case, in person, I would have reached out. Locked in a screen, all I could do was say "I'm sorry" and watch their grief.

In time, it became a routine to "see" patients on my laptop in my back room, where the sun beats on my back through the skylights and the windows overlook the maples, elms and hemlocks in the ravine below. Between the office assistants, genetics counsellors and me, after a steep learning curve, we developed a workflow.

Results and consult requests were scanned, encrypted and emailed to my inbox. And our patients were grateful for having access to genetic consultation and counselling despite the lockdown. Most were comfortable with the technology; many were genuinely grateful for not having to travel or visit a hospital and be exposed to Covid-19, and soon my misgivings dispersed.

And so it went. Summer 2020 arrived and some restrictions were lifted, but we continued to see patients by telemedicine only. One day in mid-June, I sat in my hospital office, unable to concentrate on anything, doomscrolling Covid-19 news. I was on call that week so I had filled out the online screening questionnaire and, face masked and hands sanitized, presented my proof of negative screening to the security guard at the hospital entrance. That afternoon, my pager buzzed me out of the internet rabbit hole: the neonatal intensive care unit had just admitted a newborn with complex and unexpected anomalies. Excited to be needed, I climbed the stairs two at a time, stomped down the fourth-floor corridor and scanned my ID at the NICU secure double door. At the nurses' desk, I asked where the new baby was. The receptionist looked at me askance and said that the baby was still in the labour and delivery suite across the hall. I turned on my heel and dashed out. When the door to the resuscitation room whooshed open, five blue-masked faces turned toward me and I realized with horror that I didn't have mine on.

"Oh, shit," I said. "My mask! I just got up from my desk and ran when you called." I didn't want them to think that I was flouting the rules.

A nurse wordlessly pointed to the blue box with masks on the wall and I grabbed one.

"You got here faster than the surgeons," the NICU attending joked.

"Minus the mask, that is," I mumbled as I struggled to slip the elastics behind my ears.

The speed with which I had run to see my first real live patient in weeks should have been a clue. All the technology, becoming proficient and efficient at electronic charting, improving my typing and retrieval skills, my whole techie evolution to catch up with the twenty-first century, had nothing on interacting with people: examining the baby under the warmth of the overhead heater, feeling the soft newborn skin under my fingertips, smelling that smell of freshly laundered hospital linens and a whiff of betaine. Discussing the findings with the neonatologist and neurosurgeon and the urologist, arriving at the most appropriate tests and management plan. Seeing the half-faces of my colleagues, eyes and bodies intent on giving the sick newborn the best care possible, and later, talking to the parents, seeing their eyes darting around the NICU machinery, their bodies rigid. Having them all in my physical space – the examining room or the office or the NICU bedside – was what I was used to and that comfort wasn't there during video calls. For a while, I hadn't realized that something was missing – even seeing the NICU baby didn't drive the point home. It took another month of minimal human contact; it took Sophie and her twin sister, seeing her little face as she said, "Thank you, Doctor," hearing the soft, childish voice; being able to read her abdomen with my fingers, the softness and warmth of her belly, for the scales to fall off my eyes five months after lockdown began.

But something else was afoot, something I realized much later. With the outpatient clinic closed, with genetic care delivered by video, I must have subconsciously begun to worry whether I was superfluous. If the clinic could remain closed for five months without my seeing patients in real life, what did that say about genetic consultations? Definitely not an emergency, although in the prenatal clinic it could be. I felt useless, unneeded. All my training, all my experience – it seemed colleagues and patients were doing well enough without them. Was I even needed? This gnawed on me, like a worm chewing underneath the skin of a seemingly healthy apple.

Beginning on November 1, 2020, McMaster Children's Hospital returned to pre-Covid patient volume, except that this time it would be combined in-person and virtual visits. The first Monday was a hospital holiday that I had completely forgotten about – yet another day of isolation. The next day, two of my telemedicine patients cancelled. Unmoored, unoccupied, feeling painfully alone, I lumbered down two flights of stairs to the genetics outpatient clinic to find it deserted – only the receptionist and the genetic assistant sat sequestered in their Plexiglas-delineated spaces, all of the five genetic counsellors working from home. I heard no voices, no kids zoomed around the waiting room, even the TV was turned off. All the surfaces in the examining rooms disinfected, all the doors closed – a ghost clinic. My steps echoing in the empty hallway, I reeled with loneliness and returned to my office an emotional wreck.

Wednesday, when I arrived for my awaited in-person clinic, I was delighted to find a resident waiting for me. Thankful for her company and interest, I taught her how to examine for arachnodactyly and increased arm span. I listed the Ghent diagnostic criteria for Marfan syndrome and the Beighton hypermobility score for Ehlers-Danlos syndrome, and the differences between the two conditions. I explained the types of next-generation sequencing panels for both. The brown eyes above her mask smiled – she was engaged, enjoying our interaction. In the examining room, I got down on my hands and knees to assess a woman's heels for talus valgus. I joked with an eight-year-old about his fearsome *babcia* and told his mother that I, too, was raised by a Polish mother. The boy's parents and the resident laughed – I had told a joke, the first in months. It helped that I diagnosed the boy with childhood-related hypermobility and not Ehlers-Danlos syndrome and that I was able to reassure his parents. It didn't matter that we all wore masks or that we couldn't shake hands – we were together.

Buzzed by the human contact and by the reciprocity of social interactions, I happily plowed through a stack of charts, and read, edited and electronically signed fifteen letters, but being in the land of screens didn't bother me anymore. I chatted with the genetic counsellors – all of them were in the office that day while until that day, for months, I had only seen them on Zoom calls for our semi-weekly meetings. I commiserated with the receptionist overwhelmed by a recent spate of reschedulings. I thanked the genetic assistant for the months of organizing the blood tests and send-outs for our video patients.

That afternoon, I had a standing Zoom call with a scientific program committee organizing the Canadian College of Medical Geneticists' annual meeting – I was its chair. I knew that I had patients scheduled for the afternoon as well, but I thought those were video patients with the counsellors and I was planning to juggle them around my meeting. As I was slipping on my coat to rush home for lunch before the afternoon started, a counsellor asked me where I was going. "You have a combo clinic," she said. I stopped in my tracks. Combo clinics were those where a genetic counsellor interviews the patients and I examine them and, together, we propose a plan of testing and management. There were six such patients scheduled for that afternoon.

Normally, such a change in plans would have irritated me, but that day my shoulders slid down my back and I smiled. I would work around my Zoom meeting and see the – actual, breathing, living – patients in the clinic. Ask them about their concerns and worries. Hear their inflection and diction and word choices. See their bodies react. Examine them for hypermobility of joints, stretch marks, flat feet; listen to their hearts and lungs; and palpate their pulse on the inside of their wrists. Offer a diagnosis or reassurance or plan genetic testing. I wriggled my arms out of my coat, hung it on the door and walked down the corridor to the clinic room. I knocked on the door and slipped inside.

"Hi, I'm Dr. Nowaczyk," I said to the middle-aged woman sitting on the examining table. "It's so nice to meet you."

ACKNOWLEDGEMENTS

For their support and expertise, I thank the editors of the following publications in which several of these essays first appeared, often in slightly different form:

"Marrow Memory" in *Grain* magazine
"Matisse, the Sea and Me" in *The Antigonish Review*
"Knitting Class" in *Geist* magazine
"*Ad infinitum*" in *Canadian Medical Association Journal*, reprinted in *Geist* magazine
"You, in Translation" in *Event* magazine, reprinted in *Wherever I Find Myself: Stories by Canadian Immigrant Women*, edited by Miriam Matejova
"Mother-Daughter Phrasebook: Definitions" in *Humber Literary Review*
"Almost Perfect" in *The Examined Life Journal*
"Sensorium" in *Fourth Genre*

"Metanoias" in *Geist* magazine
"What Was Missing" in *The COVID Journals: Health Care Workers Write the Pandemic*, edited by Shane Neilson, Sarah Fraser and Arundhati Dhara

My dear writing friends Christine Miscione, Sarah McNiven and Kasia Jaronczyk have been cheering me on for over a decade – a huge thank you. To my online nonfiction writing pals, Jane Silcott, Sara Graefe and Alyson Soko, gratitude for your unwavering encouragement and friendship during the pandemic and after.

I have had the incredible good fortune to take part in several online writing courses led by the incomparable Nicole Breit who gently coaxes out the most difficult stories. Thank you, Nicole, for your kindness and wisdom.

To my esteemed MFA instructors at the University of British Columbia, Kevin Chong and Wayne Grady, my thesis supervisor, a big thank you for your amazing teaching of all things nonfiction.

At Wolsak and Wynn, I would like to thank Noelle Allen, an editor extraordinaire; Ashley Hisson, for help and support above and beyond her official roles; Jen Rawlinson who designed the kick-ass cover in my favourite colours; and Megan Beadle for her keen-eyed copy edits.

And finally and most importantly, my brilliant and long-suffering husband, Jimm Douketis, and my amazing sons, Jack and Luke – you three are my world.

Born in Poland, Margaret Nowaczyk is a pediatric clinical geneticist and a professor at McMaster University and DeGroote School of Medicine and the author of *Chasing Zebras. A Memoir of Genetics, Mental Health and Writing*. Her short stories and essays have appeared in Canadian, Polish and American literary magazines and anthologies. She lives in Hamilton, ON, with her husband and two sons. Visit her website at www.margaretnowaczyk.ca.